T0263525

PIPING MATERIALS SELECTION AND APPLICATIONS

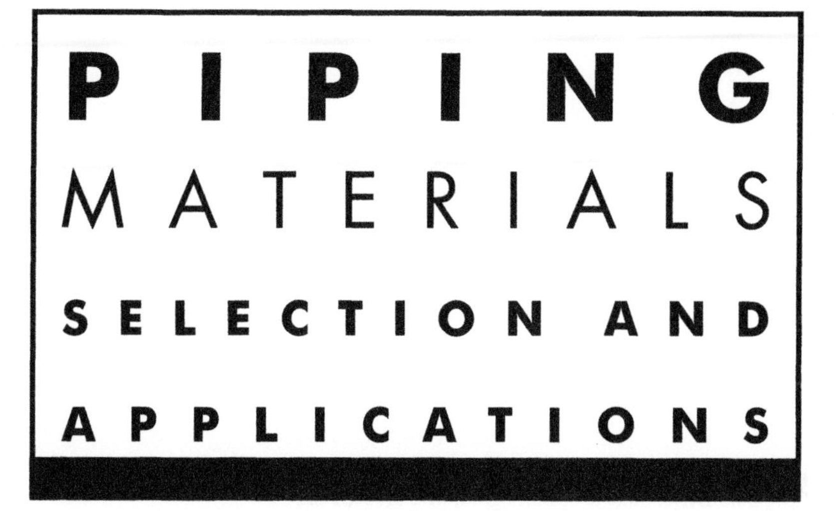

PIPING MATERIALS SELECTION AND APPLICATIONS

By

PETER SMITH

AMSTERDAM • BOSTON • HEIDELBERG • LONDON • NEWYORK • OXFORD
PARIS • SANDIEGO • SAN FRANCISCO • SINGAPORE • SYDNEY • TOKYO

Gulf Professional Publishing is an imprint of Elsevier

Gulf Professional Publishing is an imprint of Elsevier
30 Corporate Drive, Suite 400, Burlington, MA 01803, USA
Linacre House, Jordan Hill, Oxford OX2 8DP, UK

Library of Congress Cataloging-in-Publication Data
Application submitted.

British Library Cataloguing-in-Publication Data
A catalogue record for this book is available from the British Library.

ISBN: 0-7506-7743-0

For information on all Gulf Professional Publishing
publications visit our Web site at www.books.elsevier.com
Transferred to Digital Printing 2011

CONTENTS

PREFACE

The Piping Material Selection Guide for Process Systems, as the title states, is a guide for the piping engineer who is faced with the challenge of choosing the correct piping materials of construction.

The list of codes and standards in ASME B31.3 that apply to process plant design is huge, and it is impossible to cover them all in one book. Instead I use ASME B31.3 as the basic construction code, and I briefly touch on the most significant codes and standards applicable to the design of the plant.

The EPC contractor is responsible for having all the necessary codes and standards available at all times during the design, construction, and commissioning of the plant. These standards must be the final reference point, and the objective of this book is to guide the piping engineer to that point.

Although the function of a piping material engineer is driven by code and specifications, there is no substitute for all-around experience. This can be gained from several areas: the design office, a manufacturer's facility, the fabrication yard, as well as the job site. Exposure to as many of these facets of the process industry as possible is beneficial to the growth of an engineer's professional development. Each sector has its own characteristics, and knowledge of one aids the comprehension the others.

The design office is where the project evolves and is engineered and developed on paper. A manufacturer's facility is were the numerous individual components essential for construction of the project are built. For piping this includes pipe, fittings, flanges, valves, bolts, gaskets, and the like. In the fabrication yard, the welded piping components are "spooled" up for transportation to the job site. At the job site, paper and hardware come together and final fabrication and erection take place. The various piping systems are commissioned, and the project is brought to its conclusion and finally handed over to the client's operators. All these phases of a project are equally important, and it is very important that the engineer understand the challenges that arise in these very different environments.

The Piping Material Selection Guide for the Process Systems is written to be useful to all piping engineers and designers involved in the design, construction, and commissioning of oil, gas, and petrochemical facilities. However, it is primarily aimed at the piping material engineer, the individual responsible for the selection and the specifying of piping material for process facilities.

Piping engineering and the materials used in the construction of piping systems is a huge subject. It is virtually impossible to cover all aspects of it in depth in one volume. In this book, I try to cover the most important areas and introduce the reader to the fundamentals of the specific subjects. I suggest readers skim through the pages to gain a familiarity with the topics covered. I have introduced each subject and then linked it with text and technical data. I limit my use of opinions and concentrate on mandatory statements that are set out in the design codes. These standards must be met or improved on.

Most of the individuals I have worked with have developed their skills by working with fellow engineers who imparted their knowledge to the uninitiated. The ingredients that go into making a good engineer are not fully taught in schools, colleges, or universities, but by experience gained listening to more-knowledgeable colleagues, absorbing information, and through personal research.

To be a complete engineer, it is essential not only to have knowledge but to share this knowledge with fellow piping engineers and other colleagues. A piping material engineer's role is driven by codes, standards, technical data, and catalogued information. When asked a question I believe that, if possible, the answer should be supported with a copy from the relevant source of information. This allows recipients to file the information, makes them more confident, and protects the piping material engineer. It is a small action that pays big dividends.

Despite several excellent textbooks on piping design and piping stress, I know of none that specializes in piping materials. It is not the intention of this book to explain the geometry of the numerous piping components and how their final shape is computed. All the piping components discussed in this book are covered by strict design codes or recognized manufacturers' standards. Their dimensions are carefully calculated and unlikely to change dramatically in the near or distant future. Indeed, most have remained the same dimensionally for several decades and longer.

Piping engineering is not rocket science. As a fellow engineer, not a piping specialist, once said, "I thought that the Romans sorted piping out." Not true, but I see where my colleague was coming from. The piping content of a project is generally the largest of all the disciplines in material value, engineering, and construction personnel. Piping engineering also creates large volumes of paper in the form of drawings, specifications, and support documents. What it lacks in technical complexity it more than makes up for by the volumes of paperwork, which seem to increase each year.

So, to conclude, although piping may not advance as quickly as other disciplines, such as instrumentation and electrical, which are driven greatly by vendors and technology, piping does not stand still. New materials are always being developed, as well as fresh methods of manufacturing and new designs, that constantly fine-tune what we inherited from our friends the Romans.

If this book does not completely answer your questions, I feel sure that it will guide you in the right direction.

Peter Smith
Fano, Italy
June 2004

1

THE PIPING MATERIAL ENGINEER

1. WHAT IS A PIPING MATERIAL ENGINEER?

This chapter explains briefly the role of the piping engineer, who is responsible for the quality of piping material, fabrication, testing, and inspection in a project and the major activities such engineers are expected to perform. This individual can be employed by either the EPC (engineering, procurement, and construction) contractor or the operator/ end user.

1.1. Job Title

The piping engineer, the individual responsible for creating the project piping classes and the numerous piping specifications necessary to fabricate, test, insulate, and paint the piping systems, is titled either the *piping material engineer* or the *piping spec*(ification) *writer*.

1.2. Job Scope

Whatever the title, the piping material engineer (PME) is a very important person within the Piping Design Group and should be dedicated to a project from the bid stage until the design phase has been completed. He or she should also be available during construction and through to mechanical completion.

The lead piping material engineer, the individual responsible for all piping engineering functions, usually reports directly to the project lead piping engineer, and depending on the size of the project, the lead piping material engineer may be assisted by a number of suitably qualified piping material engineers especially during the peak period of the project. This peak period is early in the job, while the piping classes are being developed and the first bulk inquiry requisitions are sent out to vendors.

1.3. The Piping Material Engineer's Responsibilities

The piping material engineer's responsibilities vary from company to company. Here is a list of typical functions that he or she is expected to perform:

- Develop the project piping classes for all process and utility services.
- Write specifications for fabrication, shop and field testing, insulation, and painting.
- Create and maintain all data sheets for process and utility valves.
- Create a list of piping specials, such as hoses and hose couplings, steam traps, interlocks.
- Create and maintain data sheets for these piping special (SP) items.
- Assemble a piping material requisition with all additional documents.
- Review offers from vendors and create a technical bid evaluation.
- Make a technical recommendation.
- After placement of a purchase order, review and approve documentation from vendors related to piping components.
- When required, visit the vendor's premises to attend kickoff meetings, the testing of piping components, or clarification meetings.
- Liaise with the following departments: Piping Design and Stress, Process, Instrumentation, Vessels, Mechanical, Structural, Procurement, Material Control.

1.4. Qualities of an Engineer

Not only is it essential that a piping material engineer be experienced in several piping sectors, such as design, construction, and stress, he or she must also be a good communicator, to guarantee that everyone in the piping group is aware of the materials of construction that can be used for piping systems.

The PME must also have a basic understanding of other disciplines having interface with the piping, such as mechanical, process, instrumentation, and structural engineering. He or she should also be aware of the corrosion characteristics of piping material and welding processes necessary for the fabrication of piping systems. Both corrosion and welding engineering are specialist subjects, and if the PME has any doubts, he or she must turn to a specialist engineer for advice.

1.5. Experience

There is no substitute for experience, and the piping material engineer should have strengths in several sectors and be confident with a number of others disciplines, to enable the individual to arrive at a suitable conclusion when selecting material for piping systems.

Strong areas should include piping design layout and process requirements. Familiar areas should include the following:

- Corrosion.
- Welding.
- Piping stress.
- Static equipment.
- Rotating equipment.
- Instruments.

2. PIPING MATERIAL ENGINEER'S ACTIVITIES

Outlined here are the principal activities of a piping material engineer. These are listed in chronological order as they would arise as a project develops from preliminary to detailed design.

2.1. Development of the Project Piping Classes

All process plants have of two types of principal piping systems: process (primary and secondary) piping systems and utility piping systems.

Process piping systems are the arteries of a process plant. They receive the feedstock, carry the product through the various items of process equipment for treatment, and finally deliver the refined fluid to the battery limits for transportation to the next facility for further refinement. Process piping systems can be further divided into primary process, which is the main process flow, and secondary process, which applies to the various recycling systems.

Utility piping systems are no less important. They are there to support the primary process, falling into three groups:

- Support—instrument air, cooling water, steam.
- Maintenance—plant air, nitrogen.
- Protection—foam and firewater.

There are other utility services such as drinking water.

Piping Classes. Each piping system is allocated a piping class, which lists all the components required to construct the piping. A piping class includes the following:

- Process design conditions.
- Corrosion allowance.
- List of piping components.
- Branch table.
- Special assemblies.
- Support notes.

Both process and utility piping systems operate at various temperatures and pressures, and the following must be analyzed:

- Fluid type—corrosivity, toxicity, viscosity.
- Temperature range.
- Pressure range.
- Size range.

- Method of joining.
- Corrosion allowance.

After analyzing these characteristics, process and utility piping systems can be grouped into autonomous piping classes. This allows piping systems that share fundamental characteristics (pipe size range, pressure and temperature limits, and method of joining) to be classified together.

This standardization or optimization has benefits in the procurement, inspection, and construction phases of the project. Too little optimization increases the number of piping classes, making the paperwork at all stages of the project difficult to handle and leading to confusion, resulting in mistakes. Too much optimization reduces the number of piping classes, however, as the piping class must satisfy the characteristics of the most severe service and use the most expensive material. This means that less-severe services are constructed using more-expensive material, because the piping class is "overspecified." It is the responsibility of the piping material engineer to fine-tune this optimization to the benefit the project.

A typical oil and gas separation process plant may have 10 process piping classes and a similar number of utility piping classes. More-complex petrochemical facilities require a greater number of piping classes to cover the various process streams and their numerous temperature and pressure ranges. It is not uncommon for process plants such as these to have in excess of 50 process and piping classes.

2.2. Writing Specifications for Fabrication, Shop and Field Testing, Insulation, and Painting

It is pointless to specify the correct materials of construction if the pipes are fabricated and erected by poorly qualified labor, using bad construction methods and inadequate testing inspection, insulation, and painting.

The piping material engineer is responsible for writing project-specific narratives covering these various activities to guarantee that they meet industry standards and satisfy the client's requirements. No two projects are the same; however, many projects are very similar and most EPC companies have corporate specifications that cover these subjects.

2.3. Creating All Data Sheets for Process and Utility Valves

All valves used within a process plant must have a dedicated valve data sheet (VDS). This document is, effectively, the passport for the component, and it must detail the size range, pressure rating, design temperature, materials of construction, testing and inspection procedures and quote all the necessary design codes relating to the valve.

This VDS is essential for the efficient procurement and the possible future maintenance of the valve.

2.4. Creating a List of Piping Specials and Data Sheets

A piping system generally comprises common components such as pipe, fittings, and valves; however, less common piping items may be required, such as strainers, hoses and hose couplings, steam traps, or interlocks. This second group, called *piping specials*, must carry an SP number as an identifying tag.

The piping material engineer must create and maintain a list of SP numbers that makes the "special" unique, based on type, material, size, and rating. This means that there could be several 2 in. ASME 150, ASTM A105 body strainers with the same mesh.

As with valves, each piping special must have its own data sheet, to guarantee speedy procurement and future maintenance.

2.5. Assembling Piping Material Requisition with All Additional Documents

When all the piping specifications have been defined and initial quantities identified by the Material Take-off Group, the piping material engineer is responsible for assembling the requisition packages.

The Procurement Department will break the piping requirements into several requisitions, so that inquiry requisitions can be sent out to manufacturers or dealers that specialize in that particular group of piping components.

- Pipe (seamless and welded)—carbon and stainless steel.
- Pipe (exotic)—Inconel, Monel, titanium.

- Pipe fittings (seamless and welded)—carbon and stainless steel.
- Valves gate/globe/check (small bore, 1½ in. and below)—carbon and stainless steel.
- Valves gate/globe/check (2 in. and above)—carbon and stainless steel.
- Ball valves (all sizes)—carbon and stainless steel.
- Special valves (all sizes)—non-slam-check valves, butterfly valves.
- Stud bolting—all materials.
- Gaskets—flat, spiral wound, ring type.
- Special piping items (SPs)—strainers, hoses, hose couplings, sight glasses, interlocks, and the like.

To get competitive bids, inquiries will go out to several manufacturers for each group of piping components, and they will be invited to offer their best price to satisfy the scope of supply for the requisition. This includes not only supplying the item but also testing, certification, marking, packing, and if required, shipment to the site.

2.6. Reviewing Offers from Vendors and Create a Technical Bid Evaluation

Many clients have an "approved bidders list," which is a selection of vendors considered suitable to supply material to the company. This bidders list is based on a track record on the client's previous projects and reliable recommendations.

Prospective vendors are given a date by which they must submit a price that covers the scope of supplies laid out in the requisition. The number of vendors invited to tender a bid varies, based on the size and complexity of the specific requisition.

To create a competitive environment, a short list of between three and six suitable vendors should be considered, and it is essential that these vendors think that, at all times, they are bidding against other competitors. Even if, sometimes, vendors drop out and it becomes a "one-horse race" for commercial and technical reasons, all vendors must think that they are not bidding alone.

All vendors that deliver feasible bids should be evaluated, and it is the responsibility of the piping material engineer to bring all vendors to the same starting line and ensure that they are all offering material that meets the specifications and they are "technically acceptable," sometimes called "fit for purpose."

Some vendors will find it difficult, for commercial or technical reasons, to meet the requirements of the requisition. These vendors are deemed technically unacceptable and not considered further in the evaluation.

The piping material engineer, during this evaluation, creates a bid tabulation spreadsheet to illustrate and technically evaluate all vendors invited to submit a bid for the requisition.

The tabulation lists the complete technical requirements for each item on the requisition and evaluates each vendor to determine if it is technically acceptable.

Technical requirements include not only the materials of construction and design codes but also testing, certification, and painting. Non-technical areas also are covered by the piping material engineer, such as marking and packing. The delivery, required on site (ROS) date, is supplied by the Material Control Group as part of the final commercial negotiations.

The Procurement Department is responsible for all commercial and logistical aspects of the requisition, and the Project Services Group determines the ROS date and the delivery location. It is pointless to award an order to a manufacturer that is technically acceptable and commercially the cheapest if its delivery dates do not meet the construction schedule.

When this technical bid evaluation (TBE) or technical bid analysis (TBA) is complete, with all technically acceptable vendors identified, then it is turned over to the Procurement Department, which enters into negotiations with those vendors that can satisfy the project's technical and logistical requirements.

After negotiations, a vendor is selected that is both technically acceptable and comes up with the most competitive commercial/logistical offer. The successful vendor is not necessarily the cheapest but the one that Procurement feels most confident with in all areas. What initially looks to be the cheapest might, at the end of the day, prove more expensive.

2.7. After Placement of a Purchase Order, Reviewing and Approving Documentation Related to All Piping Components

The importance of vendor documentation after placement of an order must not be underestimated. It is the vendor's responsibility to supply support documentation and drawings to back up the material it is supplying. This documentation includes an inspection and testing plan, general arrangement drawings, material certification, test certificates, and production schedules.

All this documentation must be reviewed by the piping material engineer, approved and signed off, before final payment can be released to the vendor for the supply of the material.

2.8. Vendor Visits

The piping material engineer may be required to visit the vendor's premises to witness the testing of piping components or attend clarification meetings.

Certain piping items are more complex than others, either because of their chemical composition and supplementary requirements or their design, size, or pressure rating. In these cases, the relevant purchase order requires a greater deal of attention from the piping material engineer to ensure that no complications result in incorrect materials being supplied or an unnecessary production delay.

To avoid this, the following additional activities should be seriously considered:

- A bid clarification meeting to guarantee that the prospective vendor fully understands the requisition and associated specification.
- After the order has been placed, a preinspection meeting to discuss production, inspection, and quality control.
- Placing the requisition engineer in the vendor's facilities during critical manufacturing phases of the job to ensure that the specifications are understood.
- Placing an inspector in the vendor's facilities, who is responsible for the inspection and testing of the order and coordinates with the piping material engineer in the home office to guarantee that the specifications are understood and being applied.

The first two are low-cost activities and should be a formality for most purchase orders, the last two are more-expensive activities and should be considered based on the complexity of the order or the need for long lead items.

No two requisitions are the same, and a relatively simple order with a new and untried vendor may require more consideration than a complex order with a vendor that is a known quantity. The decision to make vendor visits also relates to the size of the inspection budget, which might not be significant enough to support "on-premises" personnel during the manufacturing phase.

Remember that if the wrong material arrives on site, then the replacement cost and the construction delay will be many times the cost of on-premises supervision.

If the items concerned are custom-made for the project or they have long lead times (three months or more), then on-premises supervision should be seriously considered.

2.9. Bids for New Projects

All the preceding are project-related activities; however, the piping material engineer may also be required to work on bids that the company has been invited to tender by clients. This is preliminary engineering, but the work produced should be accurate, based on the information provided in a brief form the client. The usual activities are preliminary piping classes, basic valve data sheets and a set of specifications for construction, inspection, and painting.

A piping material engineer will either be part of a project task force dedicated to one job or part of a corporate group working on several projects, all in different stages of completion. Of these two options, the most preferable is the former, because it allows the PME to become more familiar with the project as it develops.

The role of a piping material engineer is diverse and rewarding, and there is always something new to learn. A project may have the same client, the same process, and be in the same geographical location, but because of different personnel, a different budget, purchasing in a different market, or a string of other factors, different jobs have their own idiosyncrasies. Each one is different.

The knowledge you learn, whether technical or logistical, can be used again, so it is important that you maintain your own files, either digital or hard copies, preferably both.

Whether you work for one company for 30 years or 30 companies for 1 year, you will find that the role of PME is respected within the discipline and throughout the project.

As a function, it is no more important than the piping layout or piping stress engineer; however, its importance must not be underestimated. The pipe can be laid out in several different routings, but if the material of construction is wrong, then all the pipe routes are wrong, because the material is "out of spec."

2

PROCESS INDUSTRY CODES AND STANDARDS

1. INTRODUCTION

Process plants designed and constructed to the ASME B31.3 code also rely on the standardization of the components used for piping systems and the method of process plant fabrication and construction.

There are numerous standards, many of which are interrelated, and they must be referred and adhered to by design engineers and manufacturers in the process industry. These standards cover the following:

- Material—chemical composition, mechanical requirements, heat treatment, etc.
- Dimensions—general dimensions and tolerances.
- Fabrication codes—welding, threading.

Standards covering the preceding were drawn up by the following major engineering bodies:

- American Petroleum Institute (API).
- American Society for Testing and Materials (ASTM).
- American Water Works Association (AWWA).
- American Welding Society (AWS).

- Manufacturers Standardization Society (MSS).
- National Association of Corrosion Engineers (NACE).
- Society of Automotive Engineers (SAE).

Periodically, these standards are updated to bring them in line with the latest industry practices. Most of the standards have been in circulation for a number of years, and the changes are rarely dramatic; however, such changes must be incorporated into the design. It is essential that the latest revision is the final reference point.

Other countries publish comprehensive standards containing data on material, dimensions of components, and construction procedures; however, for the purpose of this book, the standards mentioned previously are referenced.

American standards are not superior to other national standards, but they are the ones most commonly used in the process industry. They are based on a long track record with a very low failure rate, so there is a high degree of confidence in these publications. Always refer to the latest edition of the relevant standards, and if necessary, make sure your company's library holds the most current version.

The remainder of this chapter is devoted to a listing of the most-used specifications current at the time of writing (July 2004). Please refer to the latest revision in the event of any of the specifications becoming updated.

2. AMERICAN PETROLEUM INSTITUTE

API Spec 5B. *Specification for Threading, Gauging and Thread Inspection of Casing, Tubing and Line Pipe Threads*, 14th edition, 1996.

API Spec 5L. *Specification for Line Pipe*, 42nd edition, 2000.

API Spec 6A. *Specification for Wellhead and Christmas Tree Equipment*, 18th edition, 2002.

API Bull 6AF. *Bulletin on Capabilities of API Flanges under Combinations of Load*, 2nd edition, 1995.

API TR 6AF1. *Temperature Derating of API Flanges under Combination of Loading*, 2nd edition, 1998.

API TR 6AF2. *Bulletin on Capabilities of API Integral Flanges under Combination of Loading*, 2nd edition, 1999.

API Spec 6D. *Specification for Pipeline Valves*, 22nd edition, 2002.

API Spec 6FA. *Specification for Fire Test Valves*, 3rd edition, 1999.

ANSI/API Spec 6FB. *Fire Test for End Connections*, 3rd edition, 1998.
API Spec 6FC. *Fire Test for Valve with Automatic Backseats*, 3rd edition, 1999.
API Spec 6FD. *Specification for Fire Test For Check Valves*, 1995.
ANSI/API RP 574. *Inspection Practices for Piping System Components*, 2nd edition, 1998.
ANSI/API Std 589. *Fire Test for Evaluation of Valve Stem Packing*, 2nd edition, 1998.
ANSI/API RP 591. *Use Acceptance of Refinery Valves*, 2nd edition, 1998.
API Std 594. *Check Valves—Water and Wafer-Lug and Double Flanged Type*, 5th edition, 1997.
API Std 598. *Valve Inspection and Testing*, 7th edition, 1996.
API Std 599. *Metal Plug Valves Flanged and Welding Ends*, 5th edition, 2002.
API Std 600. *Bolted Bonnet Steel Gate Valves for Petroleum and Natural Gas Industries*, 11th edition, 2001.
API Std 602. *Compact Steel Gate Valves Flanged Threaded Welding and Extended Body Ends*, 7th edition, 1998.
API Std 603. *Corrosion Resistant, Bolted Bonnet Gate Valves Flanged and Butt Welding Ends*, 6th edition, 2001.
ANSI/API Std 607. *Fire Test for Soft-Seated Quarter-Turn Valves*, 4th edition, 1993.
API Std 608. *Metal Ball Valves Flange Threaded and Welding Ends*, 3rd edition, 2002.
ANSI/API Std 609. *Butterfly Valves Double Flanged, Lug and Wafer Type*, 5th edition, 1997.
ANSI/API Std 1104. *Welding of Pipelines and Related Facilities*, 19th edition, 1999.
ANSI/API RP 1110. *Pressure Testing of Liquid Petroleum Pipelines*, 4th edition, 1997.
API RP 520, Part I. *Sizing, Selection and Installation of Pressure-Relieving Devices in Refineries*, 7th edition, 2000.
API RP 520, Part II. *Sizing, Selection and Installation of Pressure-Relieving Devices in Refineries*, 4th edition, 1994.
ANSI/API RP 521. *Guide for Pressure-Relieving and Depressuring Systems*, 4th edition, 1997.
API Std 526. *Flanged Steel Safety-Relief Valves*, 4th edition, 1995.
ANSI/API Std 527. *Seat Tightness of Pressure Relief Valves*, 3rd edition, 1991.
API RP 941. *Steels for Hydrogen Service at Elevated Temperatures and Pressures in Petroleum Refineries and Petrochemical Plants*, 5th edition, 1996.

3. AMERICAN SOCIETY OF MECHANICAL ENGINEERS (ASME)

3.1. Piping and Piping Systems

B31.1 (2001), Power Piping (piping for industrial plants and marine applications). This code covers the minimum requirements for the design, materials, fabrication, erection, testing, and inspection of power and auxiliary service piping systems for electric generation stations, industrial institutional plants, and central and district heating plants. The code also covers external piping for power boilers and high-temperature, high-pressure water boilers in which steam or vapor is generated at a pressure of more than 15 psig and high-temperature water is generated at pressures exceeding 160 psig or temperatures exceeding 250°F.

B31.2 (1968), Fuel Gas Piping. This has been withdrawn as a national standard and replaced by ANSI/NFPA Z223.1, but B31.2 is still available from ASME and is a good reference for the design of gas piping systems (from the meter to the appliance).

B31.3 (2002), Process Piping. This code covers the design of chemical and petroleum plants and refineries processing chemicals and hydrocarbons, water, and steam. It contains rules for the piping typically found in petroleum refineries; chemical, pharmaceutical, textile, paper, semiconductor, and cryogenic plants; and related processing plants and terminals. The code prescribes requirements for materials and components, design, fabrication, assembly, erection, examination, inspection, and testing of piping.

This code applies to piping for all fluids, including (1) raw, intermediate, and finished chemicals; (2) petroleum products; (3) gas, steam, air, and water; (4) fluidized solids; (5) refrigerants; and (6) cryogenic fluids. Also included is piping that interconnects pieces or stages within a packaged equipment assembly.

B31.4 (2002), Pipeline Transportation Systems for Liquid Hydrocarbons and Other Liquids. This code covers the requirements for the design, materials, construction, assembly, inspection, and testing of piping transporting liquids such as crude oil, condensate, natural gasoline, natural gas liquids, liquefied petroleum gas, carbon dioxide, liquid alcohol, liquid anhydrous ammonia, and liquid petroleum products between producers' lease facilities, tank farms, natural gas processing plants, refineries, stations, ammonia plants, terminals (marine, rail, and truck), and other delivery and receiving points.

The piping consists of pipe, flanges, bolting, gaskets, valves, relief devices, fittings, and the pressure-containing parts of other piping components. It also includes hangers and supports and other equipment items necessary to prevent overstressing the pressure-containing parts. It does not include support structures such as frames of buildings and building stanchions or foundations.

The requirements for offshore pipelines are found in Chapter IX. Also included within the scope of this code are the following:

- Primary and associated auxiliary liquid petroleum and liquid anhydrous ammonia piping at pipeline terminals (marine, rail, and truck), tank farms, pump stations, pressure-reducing stations, and metering stations, including scraper traps, strainers, and prover loops.
- Storage and working tanks, including pipe-type storage fabricated from pipe and fittings and the piping interconnecting these facilities.
- Liquid petroleum and liquid anhydrous ammonia piping located on property set aside for such piping within petroleum refinery, natural gasoline, gas processing, ammonia, and bulk plants.
- Those aspects of operation and maintenance of liquid pipeline systems relating to the safety and protection of the general public, operating company personnel, environment, property, and the piping systems.

B31.5 (2001), Refrigeration Piping and Heat Transfer Components. This code prescribes requirements for the materials, design, fabrication, assembly, erection, testing, and inspection of refrigerant, heat transfer components, and secondary coolant piping for temperatures as low as −320°F (−196°C), whether erected on the premises or assembled in a factory, except as specifically excluded in the following paragraphs. Users are advised that other piping code Sections may provide requirements for refrigeration piping in their respective jurisdictions.

This code *does not apply* to the following:

- Any self-contained or unit systems subject to the requirements of Underwriters Laboratories or another nationally recognized testing laboratory.

- Water piping.
- Piping designed for external or internal gauge pressure not exceeding 15 psi (105 kPa) regardless of size.
- Pressure vessels, compressors, or pumps.

The code does include all connecting refrigerant and secondary coolant piping starting at the first joint adjacent to such apparatus.

B31.8 (1999), Gas Transmission and Distribution Piping Systems. This code covers the design, fabrication, installation, inspection, testing, and safety aspects of operation and maintenance of gas transmission and distribution systems, including gas pipelines, gas compressor stations, gas metering and regulation stations, gas mains, and service lines up to the outlet of the customers meter set assembly.

Included within the scope of this code are gas transmission and gathering pipelines, including appurtenances, installed offshore to transport gas from production facilities to onshore locations; gas storage equipment of the closed pipe type, fabricated or forged from pipe or fabricated from pipe and fittings; and gas storage lines.

B31.8S (2001–2002), Managing System Integrity of Gas Pipelines. This standard applies to on-shore pipeline systems constructed with ferrous materials that transport gas. The pipeline system comprises all parts of the physical facilities through which gas is transported, including the pipe, valves, appurtenances attached to the pipe, compressor units, metering stations, regulator stations, delivery stations, holders, and fabricated assemblies.

The principles and processes embodied in integrity management are applicable to all pipeline systems. This standard is specifically designed to provide the operator (as defined in section 13) with the information necessary to develop and implement an effective integrity management program utilizing proven industry practices and processes.

The processes and approaches within this standard are applicable to the entire pipeline system.

B31.9 (1996), Building Services Piping. This code section has rules for the piping in industrial, institutional, commercial, and public buildings and multiunit residences that does not require the range of sizes, pressures, and temperatures covered in B31.1.

This code covers the requirements for the design, materials, fabrication, installation, inspection, examination, and testing of piping systems for building services. It includes piping systems in the building or within the property limits.

B31.11 (2002), Slurry Transportation Piping Systems. The code deals with the design, construction, inspection, security requirements of slurry piping systems. It covers piping systems that transport aqueous slurries of nonhazardous materials, such as coal, mineral ores, and other solids, between a slurry processing plant and the receiving plant.

B31G (1991), Manual for Determining Remaining Strength of Corroded Pipelines. This section is a supplement to B31, Code-Pressure Piping.

3.2 American Society of Mechanical Engineers

ASME Boiler and Pressure Vessel Code Sections

I. Power Boilers
II. Materials
III.1. Division 1, Rules for Nuclear Power Plant Components
III.2. Division 2, Code for Concrete Reactor Vessels and Containments
IV. Heating Boilers
V. Nondestructive Examination
VI. Recommended Rules for the Care and Operation of Heating Boilers
VII. Recommended Guidelines for the Care of Power Boilers
VIII.1. Pressure Vessels, Division 1
VIII.2. Pressure Vessels, Division 2—Alternative Rules
IX. Welding and Brazing Qualifications
X. Fiber-Reinforced Plastic Pressure Vessels
XI. Rules for In-Service Inspection of Nuclear Power Plant Components

Code Section Titles

B16.1 (1998), Cast Iron Pipe Flanges and Flanged Fittings.
B16.3 (1998), Malleable Iron Threaded Fittings.
B16.4 (1998), Cast-Iron Threaded Fittings.
B16.5 (1996), Pipe Flanges and Flanged Fittings.
B16.9 (1993), Factory-made Wrought Steel Butt Welding Fittings.
B16.10 (2000) Face-to-Face and End-to-End Dimensions of Valves.

B16.11 (2001) Forged Steel Fittings, Socket-Welding and Threaded.
B16.12 (1998) Cast-Iron Threaded Drainage Fittings.
B16.14 (1991) Ferrous Pipe Plugs, Bushings and Locknuts with Pipe Threads.
B16.15 (1985; R1994), Cast Bronze Threaded Fittings.
B16.18 (1984; R1994), Cast Copper Alloy Solder Joint Pressure Fittings.
B16.20 (1998), Metallic Gaskets for Pipe Flanges—Ring-Joint, Spiral-Wound, and Jacketed.
B16.21 (1992), Nonmetallic Flat Gaskets for Pipe Flanges.
B16.22 (1995), Wrought Copper and Copper Alloy Solder Joint Pressure Fittings.
B16.23 (1992), Cast Copper Alloy Solder Joint Drainage Fittings (DWV Drain, Waste, and Vent).
B16.24 (1991; R1998), Cast Copper Alloy Pipe Flanges and Flanged Fittings.
B16.25 (1997), Butt Welding Ends.
B16.26 (1988), Cast Copper Alloy Fittings for Flared Copper Tubes.
B16.28 (1994), Wrought Steel Butt Welding Short Radius Elbows and Returns.
B16.29 (1994), Wrought Copper and Wrought Copper Alloy Solder Joint Drainage Fittings (DWV).
B16.33 (1990), Manually Operated Metallic Gas Valves for Use in Gas Piping Systems up to 125 psig.
B16.34 (1996), Valves—Flanged, Threaded, and Welding End.
B16.36 (1996), Orifice Flanges.
B16.38 (1985; R1994), Large Metallic Valves for Gas Distribution.
B16.39 (1986; R1998), Malleable Iron Threaded Pipe Unions.
B16.40 (1985; R1994), Manually Operated Thermoplastic Gas.
B16.42 (1998), Ductile Iron Pipe Flanges and Flanged Fittings, Classes 150 and 300.
B16.44 (1995), Manually Operated Metallic Gas Valves for Use in House Piping Systems.
B16.45 (1998), Cast Iron Fittings for Solvent Drainage Systems.
B16.47 (1996), Large Diameter Steel Flanges: NPS 26 through NPS 60.
B16.48 (1997), Steel Line Blanks.
B16.49 (2000), Factory-made Wrought Steel Butt Welding Induction Bends for Transportation and Distribution Systems.
B16.104/FCI70-2, Control Valve Seat Leakage.

4. AMERICAN SOCIETY FOR TESTING AND MATERIALS

4.1. Index of ASTM Volumes

A vast majority of the materials of construction for process and utility piping systems used within a plant are covered by ASTM specifications. Materials and their testing methods are divided into 15 sections, each section subdivided into various volumes. ASTM covers materials of construction for industries other than the petrochemical process facilities and so many of the 15 volumes are not relevant to this industry. We now list the 15 sections and the various volumes.

Section 01. Iron and Steel Products

01.01. Steel piping, tubing, fittings.
01.02. Ferrous castings, ferroalloys.
01.03. Steel—plate, sheet, strip, wire; stainless steel bar.
01.04. Steel—structural, reinforcing, pressure vessel, railway.
01.05. Steel—bars, forgings, bearing, chain, springs.
01.06. Coated steel products.
01.07. Ships and marine technology.
01.08. Fasteners, rolling element bearings.

Section 02. Nonferrous-Metal Products

02.01. Copper and copper alloys.
02.02. Aluminium and magnesium alloys.
02.03. Electrical conductors.
02.04. Nonferrous Metals—nickel, cobalt, lead, tin, zinc, cadmium, precious, reactive, refractory metals and alloys; materials for thermostats, electrical heating and resistance contacts, and connectors.
02.05. Metallic and inorganic coatings, metal powders, sintered P/M structural parts.

Section 03. Metals, Test Methods, and Analytical Procedures

03.01. Metals mechanical testing, elevated and low-temperature tests, metallography.

03.02. Wear and erosion, metal corrosion.
03.03. Nondestructive testing.
03.04. Magnetic properties.
03.05. Analytical chemistry for metals, ores, and related materials (I): E 32 to E 1724.
03.06. Analytical chemistry for metals, ores, and related materials (II): E 1763 to latest, molecular spectroscopy, surface analysis.

Sections Not Relevant. The following sections are not relevant to the petrochemical industry:

Section 04. Construction.
Section 05. Petroleum products, lubricants, and fossil fuels.
Section 07. Textiles.
Section 08. Plastics.
Section 09. Rubber.
Section 10. Electrical insulation and electronics.
Section 11. Water and environmental technology.
Section 12. Nuclear, solar, and geothermal energy.
Section 13. Medical devices and services.
Section 14. General methods and instrumentation.
Section 15. General products, chemical specialties, and end-use products.

4.2 Commonly Used ASTM Specifications

Listed next are the most-common ASTM specifications used in the construction of process plants, designed and constructed to ASME B31.3 or associated codes. These ASTM specifications are listed numerically in the volume in which they appear.

Section 01. Iron and Steel Products

01.01. Steel—Piping, Tubing, Fittings

A53/A53M-02. Standard specification for pipe—steel, black and hot-dipped, zinc-coated, welded, and seamless.
A105/A105M-02. Standard specification for carbon steel forgings for piping applications.

A106-02a. Standard specification for seamless carbon steel pipe for high-temperature service.

A134-96(2001). Standard specification for pipe—steel, electric-fusion (arc)-welded (sizes NPS 16 and over).

A135-01. Standard specification for electric-resistance-welded steel pipe.

A139-00. Standard specification for electric-fusion (arc)-welded steel pipe (NPS 4 and over).

A179/A179M-90a(2001). Standard specification for seamless cold-drawn low-carbon steel heat-exchanger and condenser tubes.

A181/A181M-01. Standard specification for carbon steel forgings, for general-purpose piping.

A182/A182M-02. Standard specification for forged or rolled alloy-steel pipe flanges, forged fittings, and valves and parts for high-temperature service.

A193/A193M-03. Standard specification for alloy-steel and stainless steel bolting materials for high-temperature service.

A194/A194M-03b. Standard specification for carbon and alloy steel nuts for bolts for high-pressure or high-temperature service or both.

A210/A210M-02. Standard specification for seamless medium-carbon steel boiler and superheater tubes.

A234/A234M-03. Standard specification for piping fittings of wrought carbon steel and alloy steel for moderate- and high-temperature service.

A268/A268M-03. Standard specification for seamless and welded ferritic and martensitic stainless steel tubing for general service.

A269-02a. Standard specification for seamless and welded austenitic stainless steel tubing for general service.

A312/A312M-03. Standard specification for seamless and welded austenitic stainless steel pipes.

A320/A320M-03. Standard specification for alloy-steel bolting materials for low-temperature service.

A333/A333M-99. Standard specification for seamless and welded steel pipe for low-temperature service.

A334/A334M-99. Standard specification for seamless and welded carbon and alloy-steel tubes for low-temperature service.

A335/A335M-03. Standard specification for seamless ferritic alloy-steel pipe for high-temperature service.

A350/A350M-02b. Standard specification for carbon and low-alloy steel forgings, requiring notch toughness testing for piping components.

A358/A358M-01. Standard specification for electric-fusion-welded austenitic chromium-nickel alloy steel pipe for high-temperature service.

A369/A369M-02. Standard specification for carbon and ferritic alloy steel forged and bored pipe for high-temperature service.
A376/A376M-02a. Standard specification for seamless austenitic steel pipe for high-temperature central-station service.
A381-96(2001). Standard specification for metal-arc-welded steel pipe for use with high-pressure transmission systems.
A403/A403M-03a. Standard specification for wrought austenitic stainless steel piping fittings.
A409/A409M-01. Standard specification for welded large-diameter austenitic steel pipe for corrosive or high-temperature service.
A420/A420M-02. Standard specification for piping fittings of wrought carbon steel and alloy steel for low-temperature service.
A437/A437M-01a. Standard specification for alloy-steel turbine-type bolting material specially heat treated for high-temperature service.
A453/A453M-02. Standard specification for high-temperature bolting materials, with expansion coefficients comparable to austenitic stainless steels.
A524-96(2001). Standard specification for seamless carbon steel pipe for atmospheric and lower temperatures.
A530/A530M-03. Standard specification for general requirements for specialized carbon and alloy steel pipe.
A587-96(2001). Standard specification for electric-resistance-welded low-carbon steel pipe for the chemical industry.
A671-96(2001). Standard specification for electric-fusion-welded steel pipe for atmospheric and lower temperatures.
A672-96(2001). Standard specification for electric-fusion-welded steel pipe for high-pressure service at moderate temperatures.
A691-98(2002). Standard specification for carbon and alloy steel pipe, electric-fusion-welded for high-pressure service at high temperatures.
A789/A789M-02a. Standard specification for seamless and welded ferritic/austenitic stainless steel tubing for general service.
A790/A790M-03. Standard specification for seamless and welded ferritic/austenitic stainless steel pipe.
A815/A815M-01a. Standard specification for wrought ferritic, ferritic/austenitic, and martensitic stainless steel piping fittings.

01.02 Ferrous Castings, Ferroalloys

A47/A47M-99. Standard specification for ferritic malleable iron castings.
A48/A48M-00. Standard specification for gray iron castings.

A126-95(2001). Standard specification for gray iron castings for valves, flanges, and pipe fittings.

A216/A216M-93(2003). Standard specification for steel castings, carbon, suitable for fusion welding, for high-temperature service.

A217/A217M-02. Standard specification for steel castings, martensitic stainless and alloy, for pressure-containing parts, suitable for high-temperature service.

A278/A278M-01. Standard specification for gray iron castings for pressure-containing parts for temperatures up to 650°F (350°C).

A351/A351M-03. Standard specification for castings, austenitic, austenitic-ferritic (duplex), for pressure-containing parts.

A352/A352M-03. Standard specification for steel castings, ferritic and martensitic, for pressure-containing parts, suitable for low-temperature service.

A395/A395M-99. Standard specification for ferritic ductile iron pressure-retaining castings for use at elevated temperatures.

A426/A426M-02. Standard specification for centrifugally cast ferritic alloy steel pipe for high-temperature service.

A451/A451M-02. Standard specification for centrifugally cast austenitic steel pipe for high-temperature service.

A487/A487M-93(2003). Standard specification for steel castings suitable for pressure service.

A494/A494M-03a. Standard specification for castings, nickel and nickel alloy.

A571/A571M-01. Standard specification for austenitic ductile iron castings for pressure-containing parts suitable for low-temperature service.

01.03 Steel—Plate, Sheet, Strip, Wire; Stainless Steel Bar

A167-99. Standard specification for stainless and heat-resisting chromium-nickel steel plate, sheet, and strip.

A240/A240M-03c. Standard specification for chromium and chromium-nickel stainless steel plate, sheet, and strip for pressure vessels and for general applications.

A263-03. Standard specification for stainless chromium steel-clad plate.

A264-03. Standard specification for stainless chromium-nickel steel-clad plate, sheet, and strip.

A265-03. Standard specification for nickel and nickel-base alloy-clad steel plate.

A479/A479M-03. Standard specification for stainless steel bars and shapes for use in boilers and other high-pressure vessels.

01.04 Steel—Structural, Reinforcing, Pressure Vessel, Railway

A20/A20M-02. Standard specification for general requirements for steel plates for pressure vessels.
A36/A36M-03a. Standard specification for carbon structural steel.
A202/A202M-03. Standard specification for pressure vessel plates, alloy steel, chromium-manganese-silicon.
A203/A203M-97(2003). Standard specification for pressure vessel plates, alloy steel, nickel.
A204/A204M-03. Standard specification for pressure vessel plates, alloy steel, molybdenum.
A285/A285M-03. Standard specification for pressure vessel plates, carbon steel, low- and intermediate-tensile strength.
A299/A299M-03e1. Standard specification for pressure vessel plates, carbon steel, manganese-silicon.
A302/A302M-03. Standard specification for pressure vessel plates, alloy steel, manganese-molybdenum and manganese-molybdenum-nickel.
A353/A353M-93(1999). Standard specification for pressure vessel plates, alloy steel, 9% nickel, double-normalized and tempered.
A387/A387M-03. Standard specification for pressure vessel plates, alloy steel, chromium-molybdenum.
A515/A515M-03. Standard specification for pressure vessel plates, carbon steel, for intermediate- and higher-temperature service.
A516/A516M-03. Standard specification for pressure vessel plates, carbon steel, for moderate- and lower-temperature service.
A537/A537M-95(2000). Standard specification for pressure vessel plates, heat-treated, carbon-manganese-silicon steel.
A553/A553M-95(2000). Standard specification for pressure vessel plates, alloy steel, quenched and tempered 8% and 9% nickel.
A645/A645M-99a. Standard specification for pressure vessel plates, 5% nickel alloy steel, specially heat treated.

01.05 Steel—Bars, Forgings, Bearings, Chains, Springs

A508/A508M-03. Standard specification for quenched and tempered vacuum-treated carbon and alloy steel forgings for pressure vessels.
A675/A675M-90a(2000). Standard specification for steel bars, carbon, hot-wrought, special quality, mechanical properties.

01.06 Coated Steel Products

A123/A123M-02. Standard specification for zinc (hot-dip galvanized) coatings on iron and steel products.
A153/A153M-03. Standard specification for zinc coating (hot-dip) on iron and steel hardware.

01.07 Ships and Marine Technology. This material is not referenced in ASME B31.3.

01.08 Fasteners; Rolling Element Bearings.

A307-03. Standard specification for carbon steel bolts and studs, 60,000 psi tensile strength.
A325-02. Standard specification for structural bolts, steel, heat-treated, 120/105 ksi minimum tensile strength.
A325M-03. Standard specification for structural bolts, steel heat-treated 830 MPa minimum tensile strength (metric).
A354-03a. Standard specification for quenched and tempered alloy steel bolts, studs, and other externally threaded fasteners.
A563-00. Standard specification for carbon and alloy steel nuts.

Section 02. Non-Ferrous Metal Products

02.01 Copper and Copper Alloys

B21/B21M-01e1. Standard specification for naval brass rod, bar, and shapes.
B42-02. Standard specification for seamless copper pipe, standard sizes.
B43-98. Standard specification for seamless red brass pipe, standard sizes.
B61-02. Standard specification for steam or valve bronze castings.
B62-02. Standard specification for composition bronze or ounce metal castings.
B68-02. Standard specification for seamless copper tube, bright annealed.
B68M-99. Standard specification for seamless copper tube, bright annealed (metric).
B75M-99. Standard specification for seamless copper tube (metric).
B75-02. Standard specification for seamless copper tube.
B88-02. Standard specification for seamless copper water tube.
B88M-99. Standard specification for seamless copper water tube (metric).
B96/B96M-01. Standard specification for copper-silicon alloy plate, sheet, strip, and rolled bar for general purposes and pressure vessels.

B98/B98M-03. Standard specification for copper-silicon alloy rod, bar, and shapes.

B148-97(2003). Standard specification for aluminum-bronze sand castings.

B150/B150M-03. Standard specification for aluminum bronze rod, bar, and shapes.

B152/B152M-00. Standard specification for copper sheet, strip, plate, and rolled bar.

B169/B169M-01. Standard specification for aluminum bronze sheet, strip, and rolled bar.

B171/B171M-99e2. Standard specification for copper-alloy plate and sheet for pressure vessels, condensers, and heat exchangers.

B187/B187M-03. Standard specification for copper, bus bar, rod, and shapes and general-purpose rod, bar, and shapes.

B280-02. Standard specification for seamless copper tube for air conditioning and refrigeration field service.

B283-99a. Standard specification for copper and copper-alloy die forgings (hot pressed).

B466/B466M-98 Standard specification for seamless copper-nickel pipe and tube.

B467-88(2003) Standard specification for welded copper-nickel pipe.

B584-00 Standard specification for copper alloy sand castings for general applications.

02.02 Aluminum and Magnesium Alloys

B26/B26M-03. Standard specification for aluminum-alloy sand castings.

B209-02a. Standard specification for aluminum and aluminum-alloy sheet and plate.

B209M-03. Standard specification for aluminum and aluminum-alloy sheet and plate (metric).

B210-02. Standard specification for aluminum and aluminum-alloy drawn seamless tubes.

B210M-02. Standard specification for aluminum and aluminum-alloy drawn seamless tubes (metric).

B211-02. Standard specification for aluminum and aluminum-alloy bar, rod, and wire.

B211M-02. Standard specification for aluminum and aluminum-alloy bar, rod, and wire (metric).

B221M-02. Standard specification for aluminum and aluminum-alloy extruded bars, rods, wire, profiles, and tubes (metric).

B221-02. Standard specification for aluminum and aluminum-alloy extruded bars, rods, wire, profiles, and tubes.
B241/B241M-02. Standard specification for aluminum and aluminum-alloy seamless pipe and seamless extruded tube.
B247-02a. Standard specification for aluminum and aluminum-alloy die forgings, hand forgings, and rolled ring forgings.
B247M-02a. Standard specification for aluminum and aluminum-alloy die forgings, hand forgings, and rolled ring forgings (metric).
B345/B345M-02. Standard specification for aluminum and aluminum-alloy seamless pipe and seamless extruded tube for gas and oil transmission and distribution piping systems.
B361-02. Standard specification for factory-made wrought aluminum and aluminum-alloy welding fittings.
B491/B491M-00. Standard specification for aluminum and aluminum-alloy extruded round tubes for general-purpose applications.

02.03 Electrical Conductors. This material is not referenced in ASME B31.3.

02.04 Nonferrous Metals—Nickel, Cobalt, Lead, Tin, Zinc, Cadmium, Precious, Reactive, Refractory Metals and Alloys; Materials for Thermostats, Electrical Heating and Resistance Contacts, and Connectors

B127-98. Standard specification for nickel-copper alloy (UNS N04400) plate, sheet, and strip.
B160-99. Standard specification for nickel rod and bar.
B161-03. Standard specification for nickel seamless pipe and tube.
B162-99. Standard specification for nickel plate, sheet, and strip.
B164-03. Standard specification for nickel-copper alloy rod, bar, and wire.
B165-93. Standard specification for nickel-copper alloy (UNS N04400)* seamless pipe and tube.
B166-01. Standard specification nickel-chromium-iron alloys (UNS N06600, N06601, N06603, N06690, N06693, N06025, and N06045) and nickel-chromium-cobalt-molybdenum alloy (UNS N06617) rod, bar, and wire.
B167-01. Standard specification for nickel-chromium-iron alloys (UNS N06600, N06601, N06603, N06690, N06693, N06025, and N06045) and nickel-chromium-cobalt-molybdenum alloy (UNS N06617) seamless pipe and tube.
B168-01. Standard specification for nickel-chromium-iron alloys (UNS N06600, N06601, N06603, N06690, N06693, N06025, and N06045) and

nickel-chromium-cobalt-molybdenum alloy (UNS N06617) plate, sheet, and strip.

B265-02. Standard specification for titanium and titanium-alloy strip, sheet, and plate.

B333-03. Standard specification for nickel-molybdenum alloy plate, sheet, and strip.

B335-03. Standard specification for nickel-molybdenum alloy rod.

B338-02. Standard specification for seamless and welded titanium and titanium-alloy tubes for condensers and heat exchangers.

B363-03. Standard specification for seamless and welded unalloyed titanium and titanium-alloy welding fittings.

B381-02. Standard specification for titanium and titanium-alloy forgings.

B407-01. Standard specification for nickel-iron-chromium alloy seamless pipe and tube.

B409-01. Standard Specification for nickel-iron-chromium alloy plate, sheet, and strip.

B435-03. Standard specification for UNS N06002, UNS N06230, UNS N12160, and UNS R30556 plate, sheet, and strip.

B443-00e1. Standard specification for nickel-chromium-molybdenum-columbium alloy (UNS N06625) and nickel-chromium-molybdenum-silicon alloy (UNS N06219) plate, sheet, and strip.

B444-03. Standard specification for nickel-chromium-molybdenum-columbium alloys (UNS N06625) and nickel-chromium-molybdenum-silicon alloy (UNS N06219) pipe and tube.

B446-03. Standard specification for nickel-chromium-molybdenum-columbium alloy (UNS N06625), nickel-chromium-molybdenum-silicon alloy (UNS N06219), and nickel-chromium-molybdenum-tungsten alloy (UNS N06650) rod and bar.

B462-02. Specification for forged or rolled UNS N06030, UNS N06022, UNS N06200, UNS N08020, UNS N08024, UNS N08026, UNS N08367, UNS N10276, UNS N10665, UNS N10675, and UNS R20033 alloy pipe flanges, forged fittings and valves and parts for corrosive high-temperature service.

B463-99. Standard specification for UNS N08020, UNS N08026, and UNS N08024 alloy plate, sheet, and strip.

B464-99. Standard specification for welded UNS N08020, UNS N08024, and UNS N08026 alloy pipe.

B493-01(2003). Standard specification for zirconium and zirconium alloy forgings.

B514-95(2002)e1. Standard specification for welded nickel-iron-chromium alloy pipe.
B517-03. Standard specification for welded nickel-chromium-iron-alloy (UNS N06600, UNS N06603, UNS N06025, and UNS N06045) pipe.
B523/B523M-02. Standard specification for seamless and welded zirconium and zirconium alloy tubes.
B550/B550M-02. Standard specification for zirconium and zirconium alloy bar and wire.
B551/B551M-02. Standard specification for zirconium and zirconium alloy strip, sheet, and plate.
B564-00a. Standard specification for nickel alloy forgings.
B574-99a. Specification for low-carbon nickel-molybdenum-chromium, low-carbon nickel-chromium-molybdenum, low-carbon nickel-molybdenum-chromium-tantalum, low-carbon nickel-chromium-molybdenum-copper, low-carbon nickel-chromium-molybdenum-tungsten alloy rod.
B575-99a. Specification for low-carbon nickel-molydbdenum-chromium, low-carbon nickel-chromium-molybdenum, low-carbon nickel-chromium-molybdenum-copper, low-carbon nickel-chromium-molybdenum-tantalum, low-carbon nickel-chromium-molybdenum-tungsten alloy plate, sheet and strip.
B619-00. Standard specification for welded nickel and nickel-cobalt alloy pipe.
B620-03. Standard specification for nickel-iron-chromium-molybdenum alloy (UNS N08320) plate, sheet, and strip.
B621-02. Standard specification for nickel-iron-chromium-molybdenum alloy (UNS N08320) rod.
B622-00. Standard specification for seamless nickel and nickel-cobalt alloy pipe and tube.
B625-99. Standard specification for UNS N08904, UNS N08925, UNS N08031, UNS N08932, UNS N08926, and UNS R20033 plate, sheet, and strip.
B658/B658M-02. Standard specification for seamless and welded zirconium and zirconium-alloy pipe.
B675-02. Standard specification for UNS N08367 welded pipe.
B688-96. Standard specification for chromium-nickel-molybdenum-iron (UNS N08366 and UNS N08367) plate, sheet, and strip.
B690-02. Standard specification for iron-nickel-chromium-molybdenum alloys (UNS N08366 and UNS N08367) seamless pipe and tube.

B705-00. Standard specification for nickel-alloy (UNS N06625, UNS N06219 and UNS N08825) welded pipe.
B725-93. Standard specification for welded nickel (UNS N02200/UNS N02201) and nickel-copper alloy (UNS N04400) pipe.
B729-00. Standard specification for seamless UNS N08020, UNS N08026, and UNS N08024 nickel-alloy pipe and tube.

Section 03. Metals, Test Methods, and Analytical Procedures

03.01 Metals Mechanical Testing, Elevated and Low-Temperature Tests, Metallography. E112-96e2. Standard test methods for determining average grain size.

03.02 Wear and Erosion, Metal Corrosion. This situation is not referenced in ASME B31.3.

03.03 Nondestructive Testing. E114-95. (2001) Standard practice for ultrasonic pulse-echo straight-beam examination by the contact method.

E125-63(2003). Standard reference photographs for magnetic particle indications on ferrous castings.
E155-00. Standard reference radiographs for inspection of aluminum and magnesium castings.
E165-02. Standard test method for liquid penetrant examination.
E186-98. Standard reference radiographs for heavy-walled (2–4½–12 in.; 51–114 mm) steel castings.
E213-02. Standard practice for ultrasonic examination of metal pipe and tubing.
E272-99. Standard reference radiographs for high-strength copper-base and nickel-copper alloy castings.
E280-98. Standard reference radiographs for heavy-walled (4½–12 in.; 114–305 mm) steel castings.
E310-99. Standard reference radiographs for tin bronze castings.
E446-98. Standard reference radiographs for steel castings up to 2 in. (51 mm) thickness.
E709-01. Standard guide for magnetic particle examination.

03.04 Magnetic Properties. Such properties are not referenced in ASME B31.3.

5. AMERICAN WELDING SOCIETY

A3.0: 2001. Standard welding terms and definitions, including terms for adhesive bonding, brazing, soldering, thermal cutting, and thermal spraying.

A5.01-93R. Filler metal procurement guidelines.

A5-ALL. Filler metal specifications series plus filler metal procurement guide.

6. AMERICAN WATER WORKS ASSOCIATION

Ductile-Iron Pipe and Fittings

C110/A21.10-03. ANSI standard for ductile-iron and gray-iron fittings, 3–48 in. (76–1219 mm), for water.

C111/A21.11-00. ANSI standard for rubber-gasket joints for ductile-iron pressure pipe and fittings.

C115/A21.15-99. ANSI standard for flanged ductile-iron pipe with ductile-iron or gray-iron threaded flanges.

C150/A21.50-02. ANSI standard for thickness design of ductile-iron pipe.

C151/A21.51-02. ANSI standard for ductile-iron pipe, centrifugally cast, for water.

Steel Pipe

C200-97. Steel water pipe—6 in. (150 mm) and larger.

C207-01. Steel pipe flanges for waterworks service—sizes 4–144 in. (100–3600 mm).

C208-01. Dimensions for fabricated steel water pipe fittings.

Concrete Pipe

C300-97. Reinforced concrete pressure pipe, steel-cylinder type.

C301-99. Prestressed concrete pressure pipe, steel-cylinder type.

C302-95. Reinforced concrete pressure pipe, no cylinder type.

Valves and Hydrants

C500-02. Metal-Seated gate valves for water supply service (includes addendum C500a-95).
C504-00. Rubber-seated butterfly valves.
C507-99. Ball valves, 6–48 in. (150–1200 mm).

Plastic Pipe

C900-97. Polyvinyl chloride (PVC) pressure pipe, and fabricated fittings, 4–12 in. (100–300 mm), for water distribution.
C950-01. Fiberglass pressure pipe.

7. MANUFACTURERS STANDARDIZATION SOCIETY

SP-6 (2001). Standard finishes for contact faces of pipe flanges and connecting-end flanges of valves and fittings.
SP-9 (2001). Spot facing for bronze, iron, and steel flanges.
SP-25 (1998). Standard marking system for valves, fittings, flanges, and unions.
SP-421 (1999). Class 150 corrosion resistant gate, globe, angle, and check valves with flanged and butt weld ends.
SP-43 (1991; R2001). Wrought stainless steel butt-welding fittings.
SP-44 (1996; R2001). Steel pipeline flanges.
SP-45 (2003). Bypass and drain connections.
SP-51 (2003). Class 150LW corrosion-resistant cast flanges and flanged fittings.
SP-53 (1999). Quality standard for steel castings and forgings for valves, flanges, and fittings and other piping components, magnetic particle exam method.
SP-54 (1999; R2002). Quality standard for steel castings for valves, flanges, and fittings and other piping components, radiographic examination method.
SP-55 (2001). Quality standard for steel castings for valves, flanges, fittings, and other piping components, visual method for evaluation of surface irregularities.

SP-58 (2002). Pipe hangers and supports—materials, design, and manufacture.
SP-60 (1999). Connecting flange joint between tapping sleeves and tapping valves.
SP-61 (2003). Pressure testing of steel valves.
SP-65 (1999). High-pressure chemical industry flanges and threaded stubs for use with lens gaskets.
SP-67 (2002). Butterfly valves.
SP-68 (1997). High-pressure butterfly valves with offset design.
SP-69 (2002). Pipe hangers and supports—selection and application.
SP-70 (1998). Cast-iron gate valves, flanged and threaded ends.
SP-71 (1997). Gray-iron swing check valves, flanged and threaded ends.
SP-72 (1999). Ball valves with flanged or butt-welding ends for general service.
SP-73 (2003). Brazing joints for copper and copper-alloy pressure fittings.
SP-75 (1998). Specification for high-test wrought butt-welding fittings.
SP-77 (1995; R2000). Guidelines for pipe support contractual relationships.
SP-78 (1998). Cast-iron plug valves, flanged and threaded ends.
SP-79 (1999a). Socket-welding reducer inserts.
SP-80 (2003). Bronze gate, globe, angle, and check valves.
SP-81 (2001). Stainless Steel, Bonnetless, Flanged Knife Gate Valves.
SP-82 (1992). Valve-pressure testing methods.
SP-83 (2001). Class 3000 steel pipe unions, socket welding and threaded.
SP-85 (2002). Cast-iron globe and angle valves, flanged and threaded ends.
SP-86 (2002). Guidelines for metric data in standards for valves, flanges, fittings and actuators.
SP-88 (1993; R2001). Diaphragm valves.
SP-89 (1998). Pipe hangers and supports—fabrication and installation practices.
SP-90 (2000). Guidelines on terminology for pipe hangers and supports.
SP-91 (1992; R1996). Guidelines for manual operation of valves.
SP-92 (1999). MSS valve user guide.
SP-93 (1999). Quality standard for steel castings and forgings for valves, flanges, and fittings and other piping components, liquid-penetrant exam method.
SP-94 (1999). Quality standard for ferritic and martensitic steel castings for valves, flanges, and fittings and other piping components, ultrasonic exam method.
SP-95 (2000). Swage (d) nipples and bull plugs.
SP-96 (2001). Guidelines on terminology for valves and fittings.

SP-97 (2001). Integrally reinforced forged branch outlet fittings—socket welding, threaded, and butt welding ends.

SP-98 (2001). Protective coatings for the interior of valves, hydrants, and fittings.

SP-99 (1994; R2001). Instrument valves.

SP-100 (2002). Qualification requirements for elastomer diaphragms for nuclear service diaphragm-type valves.

SP-101 (1989; R2001). Part-turn valve actuator attachment—flange and driving component dimensions and performance characteristics.

SP-102 (1989; R2001). Multiturn valve actuator attachment—flange and driving component dimensions and performance characteristics.

SP-103 (1995; R2000). Wrought copper and copper-alloy insert fittings for polybutylene systems.

SP-104 (2003). Wrought copper solder joint pressure fittings.

SP-105 (1996; R2001). Instrument valves for code applications.

SP-106 (2003). Cast copper-alloy flanges and flanged fittings, Class 125, 150 and 300.

SP-107 (1991; R2000). Transition union fittings for joining metal and plastic products.

SP-108 (2002). Resilient-seated cast iron-eccentric plug valves.

SP-109 (1997). Welded fabricated copper solder joint pressure fittings.

SP-110 (1996). Ball Valves threaded, socket-welding, solder joint, grooved and flared ends.

SP-111 (2001). Gray-iron and ductile-iron tapping sleeves.

SP-112 (1999). Quality standard for evaluation of cast surface finishes—visual and tactile method (this SP must be sold with a 10-surface, three-dimensional cast surface comparator, which is a necessary part of the standard).

SP-113 (2001). Connecting joint between tapping machines and tapping valves.

SP-114 (2001). Corrosion resistant pipe fittings threaded and socket welding, Class 150 and 1000.

SP-115 (1999). Excess flow valves for natural gas service.

SP-116 (2003). Service line valves and fittings for drinking water systems.

SP-117 (2002). Bellows seals for globe and gate valves.

SP-118 (2002). Compact steel globe and check valves—flanged, flangeless, threaded and welding ends (chemical and petroleum refinery service).

SP-119 (2003). Factory-made wrought belled-end socket-welding fittings.

SP-120 (2002). Flexible graphite packing system for rising-stem steel valves (design requirements).

SP-121 (1997; R2002). Qualification testing methods for stem packing for rising-stem steel valves.
SP-122 (1997). Plastic industrial ball valves.
SP-123 (1998). Nonferrous threaded and solder-joint unions for use with copper water tube.
SP-124 (2001). Fabricated tapping sleeves.
SP-125 (2000). Gray-iron and ductile-iron in-line, spring-loaded, center-guided check valves.
SP-126 (2000). Steel in-line spring-assisted center guided check valves.
SP-127 (2001). Bracing for piping systems seismic-wind-dynamic design, selection, application.
SP-129 (2003). Copper-nickel socket-welding fittings and unions.
SP-130 (2003). Bellows seals for instrument valves.

8. NATIONAL ASSOCIATION OF CORROSION ENGINEERS (NACE)

MR0175 (2003). Metals for sulfide stress cracking and stress corrosion cracking resistance in sour oilfield environments.
RP0170 (1997). Protection of austenitic stainless steels and other austenitic alloys from polythionic acid stress corrosion cracking during shutdown of refinery equipment.
RP0472 (2000). Methods and controls to prevent in-service environmental cracking of carbon steel weldments in corrosive petroleum refining environments.

9. SOCIETY OF AUTOMOTIVE ENGINEERS

SAE J513 (1999). Refrigeration tube fittings—general specifications.
SAE J514 (2001). Hydraulic tube fittings.
SAE J 518 (1993). Hydraulic flanged tube, pipe, and hose connections, four-bolt split flange type.

3

MATERIALS

1. INTRODUCTION

This chapter covers the most commonly used materials of construction for piping systems within a process plant.

The two principal international codes used for the design and construction of a process plant are ASME B31.3, Process Piping, and the ASME Boiler and Pressure Vessel Code Sections.

Generally, only materials recognized by the American Society of Mechanical Engineers (ASME) can be used as the "materials of construction" for piping systems within process plants, because they meet the requirements set out by a recognized materials testing body, like the American Society of Testing and Materials (ASTM).

There are exceptions, however; the client or end user must be satisfied that the non-ASTM materials offered are equal or superior to the ASTM material specified for the project.

The Unified Numbering System (UNS) for identifying various alloys is also quoted. This is not a specification, but in most cases, it can be cross-referenced to a specific ASTM specification.

1.1. American Society of Testing and Materials

The American Society of Testing and Materials specifications cover materials for many industries, and they are not restricted to the process sector and associated industries. Therefore, many ASTM specifications are not relevant to this book and will never be referred to by the piping engineer.

We include passages from a number of the most commonly used ASTM specifications. This gives the piping engineer an overview of the specifications and scope in one book, rather than several ASTM books, which carry specifications a piping engineer will never use.

It is essential that at the start of a project, the latest copies of all the relevant codes and standards are available to the piping engineer.

All ASTM specification identifiers carry a prefix followed by a sequential number and the year of issue; for example, A105/A105M-02, Standard Specification for Carbon Steel Forgings for Piping Applications, breaks down as follows:

A = prefix.
105 = sequential number.
M means that this specification carries metric units.
02 = 2002, the year of the latest version.
Official title = Standard Specification for Carbon Steel Forgings for Piping Applications.

The complete range of ASTM prefixes are A, B, C, D, E, F, G, PS, WK; however, the piping requirements referenced in ASME B31.3, which is considered our design "bible," call for only A, B, C, D, and E.

The requirements of an ASTM specification cover the following:

- Chemical requirements (the significant chemicals used in the production and the volumes).
- Mechanical requirements (yield, tensile strength, elongation, hardness).
- Method of manufacture.
- Heat treatment.
- Weld repairs.
- Tolerances.
- Certification.
- Markings.
- Supplementary notes.

If a material satisfies an ASTM standard, then the various characteristics of the material are known and the piping engineer can confidently use the material in a design, because the allowable stresses and the strength of the material can be predicted and its resistance against the corrosion of the process is known.

1.2. Unified Numbering System

Alloy numbering systems vary greatly from one alloy group to the next. To avoid confusion, the UNS for metals and alloys was developed. The UNS number is not a specification, because it does not refer to the method of manufacturing in which the material is supplied (e.g., pipe bar, forging, casting, plate). The UNS indicates the chemical composition of the material.

An outline of the organization of UNS designations follows:

UNS Series	Metal
A00001 to A99999	Aluminum and aluminum alloys
C00001 to C99999	Copper and copper alloys
D00001 to D99999	Specified mechanical property steels
E00001 to E99999	Rare earth and rare-earth-like metals and alloys
F00001 to F99999	Cast irons
G00001 to G99999	AISI and SAE carbon and alloy steels (except tool steels)
H00001 to H99999	AISI and SAE H-steels
J00001 to J99999	Cast steels (except tool steels)
K00001 to K99999	Miscellaneous steels and ferrous alloys
L00001 to L99999	Low-melting metals and alloys
M00001 to M99999	Miscellaneous nonferrous metals and alloys
N00001 to N99999	Nickel and nickel alloys
P00001 to P99999	Precious metals and alloys
R00001 to R99999	Reactive and refractory metals and alloys
S00001 to S99999	Heat and corrosion resistant (stainless) steels
T00001 to T99999	Tool steels, wrought and cast
W00001 to W99999	Welding filler metals
Z00001 to Z99999	Zinc and zinc alloys

In this chapter, the ASTM specification is the most common reference in the design of process plants. Extracts from a number of the most commonly used ASTM specifications are listed in the book, along with the general scope of the specification and the mechanical requirements.

For detailed information, the complete specification must be referred to and the engineering company responsible for the design of the plant must have copies of all codes and standards used as part of their contractual obligation.

1.3. Manufacturer's Standards

Several companies are responsible for inventing, developing, and manufacturing special alloys, which have advanced characteristics that allow them to be used at elevated temperatures, low temperatures, and in highly corrosive process services. In many cases, these materials were developed for the aerospace industry, and after successful application, they are now used in other sectors.

Three examples of such companies are listed below:

- Haynes International, Inc.—high-performance nickel- and cobalt-based alloys.
- Carpenter Technology Corporation—stainless steel and titanium.
- Sandvik—special alloys.

1.4. Metallic Material Equivalents

Some ASTM materials are compatible with specifications from other countries, such as BS (Britain), AFNOR (France), DIN (Germany), and JIS (Japan). If a specification from one of these other countries either meets or is superior to the ASTM specification, then it is considered a suitable alternative, if the project certifications are met.

1.5. Nonmetallic Materials

In many cases, nonmetallic materials have been developed by a major manufacturer, such as Dow Chemical, ICI, or DuPont, which holds the patent on the material. This material can officially be supplied only by the patent owner or a licensed representative.

The patent owners are responsible for material specification, which defines the chemical composition and associated mechanical characteristics. Four examples of patented materials that are commonly used in the process industry are as follows:

- Nylon, a polyamide, DuPont.
- Teflon, polytetrafluoroethylene, DuPont.
- PEEK, polyetheretherketone, ICI.
- Saran, polyvinylidene chloride, Dow.

Certain types of generic nonmetallic material covering may have several patent owners; for example, patents for PVC (polyvinyl chloride) are owned by Carina (Shell), Corvic (ICI), Vinoflex (BASF), and many others. Each of these examples has unique characteristics that fall into the range covered by the generic term PVC. To be sure of these characteristics, it is important that a material data sheet (MDS) is obtained from the manufacturer and this specification forms part of the project documentation.

2. MATERIALS SPECIFICATIONS

Listed below are extracts from the most commonly used material specifications referenced in ASME B31.3.

ASTM, A53/A53M-02 (Volume 01.01), Standard Specification for Pipe, Steel, Black and Hot-Dipped, Zinc-Coated, Welded and Seamless

Scope.

1.1 This specification covers seamless and welded black and hot-dipped galvanized steel pipe in NPS $^1/_8$ to NPS 26 (DN 6 to DN 650) for the following types and grades:

1.2.1 *Type F*—furnace-butt welded, continuous welded Grade A.

1.2.2 *Type E*—electric-resistance welded, Grades A and B.

1.2.3 *Type S*—seamless, Grades A and B.

Referenced Documents

ASTM

A90/A90M, Test Method for Weight [Mass] of Coating on Iron and Steel Articles with Zinc or Zinc-Alloy Coatings.

A370, Test Methods and Definitions for Mechanical Testing of Steel Products.

A530/A530M, Specification for General Requirements for Specialized Carbon and Alloy Steel Pipe.

A700, Practices for Packaging, Marking, and Loading Methods for Steel Products for Domestic Shipment.

A751, Test Methods, Practices, and Terminology for Chemical Analysis of Steel Products.
A865, Specification for Threaded Couplings, Steel, Black or Zinc-Coated (Galvanized) Welded or Seamless, for Use in Steel Pipe Joints.
B6, Specification for Zinc.
E29, Practice for Using Significant Digits in Test Data to Determine Conformance with Specifications.
E213, Practice for Ultrasonic Examination of Metal Pipe and Tubing.
E309, Practice for Eddy-Current Examination of Steel Tubular Products Using Magnetic Saturation.
E570, Practice for Flux Leakage Examination of Ferromagnetic Steel Tubular Products.
E1806, Practice for Sampling Steel and Iron for Determination of Chemical Composition.
ASC Acredited Standards Committee X12.

ASME

B1.20.1, Pipe Threads, General Purpose.
B36.10, Welded and Seamless Wrought Steel Pipe.

Military Standard (MIL)

STD-129, Marking for Shipment and Storage.
STD-163, Steel Mill Products Preparation for Shipment and Storage.
Fed. Std. No. 123, Marking for Shipment (Civil Agencies).
Fed. Std. No. 183, Continuous Identification Marking of Iron and Steel Products.

American Petroleum Institute (API)

5L, Specification for Line Pipe.

Methods of Manufacture. Open hearth (OH), electrofurnace (EF), basic oxygen (BO).

Chemical Requirements. Refer to ASTM A53/A53M.

Mechanical Requirements. These are extracted from ASTM A53/A53M:

Type	Grade	Manufacture	Minimum Tensile Strength, ksi (MPa)	Minimum Yield Strength, ksi (MPa)
F	A	OH, EF, BO	48.0 (330)	30.0 (205)
E, S	A	OH, EF, BO	48.0 (330)	30.0 (205)
E, S	B	OH, EF, BO	60.0 (415)	35.0 (240)

ASTM, A106-02a (Volume 1.01), Standard Specification for Seamless Carbon Steel Pipe for High-Temperature Service

Scope. This specification covers seamless carbon steel pipe for high-temperature service (Note: It is suggested that consideration be given to possible graphitization) in NPS $^1/_8$ –NPS 48 inclusive, with nominal (average) wall thickness as given in ANSI B 36.10. It is permissible to furnish pipe having other dimensions provided such pipe complies with all other requirements of this specification. Pipe ordered under this specification is suitable for bending, flanging, and similar forming operations and for welding. When the steel is to be welded, it is presupposed that a welding procedure suitable to the grade of steel and intended use or service is utilized (Note: The purpose for which the pipe is to be used should be stated in the order. Grade A rather than Grade B or Grade C is the preferred grade for close coiling or cold bending. This note is not intended to prohibit the cold bending of Grade B seamless pipe).

Referenced Documents

ASTM

A530/A530M, Specification for General Requirements for Specialized Carbon and Alloy Steel Pipe.

E213, Practice for Ultrasonic Examination of Metal Pipe and Tubing.

E309, Practice for Eddy-Current Examination of Steel Tubular Products Using Magnetic Saturation.

E381, Method of Macroetch Testing, Inspection, and Rating Steel Products, Comprising Bars, Billets, Blooms, and Forgings.

A520, Specification for Supplementary Requirements for Seamless and Electric-Resistance-Welded Carbon Steel Tubular Products for High-Temperature Service Conforming to ISO Recommendations for Boiler Construction.

E570, Practice for Flux Leakage Examination of Ferromagnetic Steel Tubular Products.

ASME

B36.10, Welded and Seamless Wrought Steel.

Methods of Manufacture. Open hearth (OH), electrofurnace (EF), basic oxygen (BO).

Chemical Requirements. Refer to from ASTM A106/A106M.

Mechanical Requirements. These are extracted from ASTM A106/A106M:

Grade	Manufacture	Minimum Tensile Strength, ksi (MPa)	Minimum Yield Strength, ksi (MPa)
A	OH, EF, BO	48.0 (330)	30.0 (205)
B	OH, EF, BO	60.0 (415)	35.0 (240)
C	OH, EF, BO	70.0 (485)	40.0 (275)

ASTM, A126-95 (2001) (Volume 01.02), Standard Specification for Gray Iron Castings for Valves, Flanges, and Pipe Fittings

Scope. This specification covers three classes of gray iron for castings intended for use as valve pressure retaining parts, pipe fittings, and flanges.

Referenced Documents

ASTM

A438, Test Method for Transverse Testing of Gray Cast Iron.
A644, Terminology Relating to Iron Castings.
E8, Test Methods for Tension Testing of Metallic Materials.
A48, Specification for Gray Iron Castings.

Sizes. Varies.

Heat Treatment. Refer to ASTM A126/A126M.

Welding Repair. For repair procedures and welder qualifications, see ASTM A488/A488M.

Chemical Requirements. Refer to ASTM A126/A126M.

Mechanical Requirements. These are extracted from ASTM A126/A126M:

Class	Minimum Tensile Strength, ksi (MPa)
A	21 (145)
B	31 (214)
C	41 (283)

ASTM, A134-96 (2001) (Volume 1.01), Standard Specification for Pipe, Steel, Electric-Fusion (Arc)-Welded (Sizes NPS 16 and Over)

Scope. This specification covers electric-fusion (arc)-welded straight seam or spiral seam steel pipe NPS 16 and over in diameter (inside or outside as specified by purchaser), with wall thicknesses up to $^3/_4$ in. (19.0 mm) inclusive. Pipe having other dimensions may be furnished provided such pipe complies with all other requirements of this specification.

Referenced Documents

ASTM

A36/A36M, Specification for Carbon Structural Steel.
A283/A283M, Specification for Low- and Intermediate-Tensile-Strength Carbon Steel Plates.
A285/A285M, Specification for Pressure Vessel Plates, Carbon Steel, Low- and Intermediate-Tensile Strength.
A370, Test Methods and Definitions for Mechanical Testing of Steel Products.
A570/A570M, Specification for Steel, Sheet and Strip, Carbon, Hot-Rolled, Structural.

ASME Boiler and Pressure Vessel Code. Section IX Welding Qualifications American National Standards Institute Standard.

ASTM, A167-99 (Volume 01.03), Standard Specification for Stainless and Heat-Resisting Chromium-Nickel Steel Plate, Sheet, and Strip

Scope. This specification covers stainless and heat-resisting chromium-nickel steel plate, sheet, and strip.

Referenced Documents

ASTM

A240/A240M, Specification for Heat-Resisting Chromium and Chromium-Nickel Stainless Steel Plate, Sheet and Strip for Pressure Vessels.
A370, Test Methods and Definitions for Mechanical Testing of Steel Products.
A480/A480M, Specification for General Requirements for Flat-Rolled Stainless and Heat-Resisting Steel Plate, Sheet, and Strip.

UNS

E527, Practice for Numbering Metals and Alloys.
J1086, Numbering Metals and Alloys.

Chemical Composition. Refer to ASTM A167.

Mechanical Requirements. These are extracted from ASTM A167:

Grade	Minimum Tensile Strength, ksi (MPa)	Minimum Yield Strength, ksi (MPa)
S31215	75.0 (515)	30.0 (205)
S30800	75.0 (515)	30.0 (205)
S30900	75.0 (515)	30.0 (205)
S31000	75.0 (515)	30.0 (205)

ASTM, A179/A179M-90a (2001) (Volume 01.01), Standard Specification for Seamless Cold-Drawn Low-Carbon Steel Heat Exchanger and Condenser Tubes

Scope. This specification covers minimum-wall-thickness, seamless cold-drawn low-carbon steel tubes for tubular heat exchangers, condensers, and similar heat transfer apparatus. The tubes are $\frac{1}{8}$–3 in. (3.2–76.2 mm), inclusive, in outside diameter. Note: Tubing smaller in outside diameter and having a thinner wall than indicated in this specification is available. Mechanical property requirements do not apply to tubing smaller than $\frac{1}{8}$ in. (3.2 mm) in outside diameter or with a wall thickness under 0.015 in. (0.4 mm).

Referenced Document

ASTM. A450/A450M Specification for General Requirements for Carbon, Ferritic Alloy, and Austenitic Alloy Steel Tubes.

Methods of Manufacture. Tubes are made by the seamless process and cold drawn.

Chemical Requirements. Refer to ASTM A179/A179M.

Mechanical Requirements. These are extracted from ASTM A179/A 179M:

Minimum Tensile Strength, ksi (MPa)	Minimum Yield Strength, ksi (MPa)
47.0 (325)	26.0 (180)

ASTM, A181/A181M-01 (Volume 01.01), Standard Specification for Carbon Steel Forgings, for General-Purpose Piping

Scope. This specification covers nonstandard as-forged fittings, valve components, and parts for general service. Forgings made to this specification are limited to a maximum weight of 10,000 lb (4540 kg). Larger forgings may be ordered to Specification A266/A266M.

Two grades of material are covered, designated Classes 60 and 70, respectively, and classified in accordance with their chemical composition and mechanical properties.

Class 60 was formerly designated Grade I, and Class 70 was formerly designated Grade II.

Referenced Documents

ASTM

A266/A266M, Specification for Carbon Steel Forgings for Pressure Vessel Components.
A788, Specification for Steel Forgings, General Requirements.
A961, Specification for Common Requirements for Steel Flanges, Forged Fittings, Valves, and Parts for Piping Applications.

Automotive Industry Action Group (AIAG). AIAG B-5 02.00, Primary Metals Identification Tag Application Standard.

Heat Treatment. Refer to ASTM A181/A181M.

Chemical Requirements. Refer ASTM A181/A181M.

Mechanical Requirements. These are extracted from ASTM A181/A181M:

	Grade	Minimum Tensile Strength, ksi (MPa)	Minimum Yield Strength, ksi (MPa)
A181	60	60 (415)	30 (250)
	70	70 (485)	36 (250)

ASTM, A182/A182M-02 (Volume 01.01), Standard Specification for Forged or Rolled Alloy-Steel Pipe Flanges, Forged Fittings, and Valves and Parts for High-Temperature Service

Scope. This specification covers forged low-alloy and stainless steel piping components for use in pressure systems. Included are flanges, fittings, valves, and similar parts to specified dimensions or dimensional standards, such as the ASME specifications referenced next.

Referenced Documents

ASTM

A234/A234M, Specification for Piping Fittings of Wrought Carbon Steel and Alloy Steel for Moderate- and High-Temperature Service.

A262, Practices for Detecting Susceptibility to Intergranular Attack in Austenitic Stainless Steels.
A275/A275M, Test Method for Magnetic Particle Examination of Steel Forgings.
A336/A336M, Specification for Alloy Steel Forgings for Pressure and High-Temperature Parts.
A370, Test Methods and Definitions for Mechanical Testing of Steel Products.
A403/A403M, Specification for Wrought Austenitic Stainless Steel Piping Fittings.
A479/A479M, Specification for Stainless Steel Bars and Shapes for Use in Boilers and Other Pressure Vessels.
A484/A484M, Specification for General Requirements for Stainless Steel Bars, Billets, and Forgings.
A739, Specification for Steel Bars, Alloy, Hot Wrought, for Elevated Temperature or Pressure-Containing Parts or Both.
A763, Practices for Detecting Susceptibility to Intergranular Attack in Ferritic Stainless Steels.
A788, Specification for Steel Forgings, General Requirements.
A961, Specification for Common Requirements for Steel Flanges, Forged Fittings, Valves, and Parts for Piping Applications.
E112, Test Methods for Determining Average Grain Size.
E165, Test Method for Liquid Penetrant Examination.
E340, Test Method for Macroetching Metals and Alloys.
Section IX, Welding Qualifications.

ASME Boiler and Pressure Vessel Code (BPV). Section IX Welding Qualifications

SFA-5.4, Specification for Corrosion-Resisting Chromium and Chromium-Nickel Steel Covered Welding Electrodes.
SFA-5.5, Specification for Low-Alloy Steel-Covered Arc-Welding Electrodes.
SFA-5.9, Specification for Corrosion-Resisting Chromium and Chromium-Nickel Steel Welding Rods and Bare Electrodes.
SFA-5.11, Specification for Nickel and Nickel-Alloy Covered Welding Electrodes.

ASME

B16.5, Dimensional Standards for Steel Pipe Flanges and Flanged Fittings.
B16.9, Steel Butt-Welding Fittings.
B16.10, Face-to-Face and End-to-End Dimensions of Ferrous Valves.

B16.11, Forged Steel Fittings, Socket Weld and Threaded.
B16.34, Valves-Flanged, Threaded and Welding End.

Sizes. Varies.

Methods of Manufacture

Low alloy—open hearth, electric furnace or basic oxygen.
Stainless steel—electric furnace, vacuum furnace, one of the former, followed by vacuum or electroslag-consumable remelting.

Heat Treatment. Refer to ASTM A182/A182M.

Chemical Requirements. Refer to ASTM A182/A182M.

Mechanical Requirements. These are extracted from ASTM A182/A182M:

Grade	Minimum Tensile Strength, ksi (MPa)	Minimum Yield Strength, ksi (MPa)
F1	70 (485)	40 (275)
F2	70 (485)	40 (275)
F5	70 (485)	40 (275)
F5a	90 (620)	65 (450)
F9	85 (585)	55 (380)
F91	85 (585)	60 (415)
F92	90 (620)	64 (440)
F911	90 (620)	64 (440)
F11 Class 1	60 (415)	30 (205)
F11 Class 2	70 (485)	40 (275)
F11 Class 3	75 (515)	45 (310)
F12 Class 1	60 (415)	32 (220)
F12 Class 2	70 (485)	40 (275)
F21	75 (515)	45 (310)
F3V, F3VCb	85–110 (585–760)	60 (415)
F22 Class 1	60 (415)	30 (205)
F22 Class 3	75 (515)	45 (310)
F22V	85–110 (585–760)	60 (415)
F23	74 (510)	58 (400)
F24	85 (585)	60 (415)
FR	63 (435)	46 (315)
F122	90 (620)	58 (400)

Grade	Minimum Tensile Strength, ksi (MPa)	Minimum Yield Strength, ksi (MPa)
F6a Class 1	70 (485)	40 (275)
F6a Class 2	85 (585)	55 (380)
F6a Class 3	110 (760)	85 (585)
F6a Class 4	130 (895)	110 (760)
F6b	110–135 (760–930)	90 (620)
F6NM	115 (790)	90 (620)
FXM-27Cb	60 (415)	35 (240)
F429	60 (415)	35 (240)
F430	60 (415)	35 (240)
F304	75 (515)	30 (205)
F304H	75 (515)	30 (205)
F304L	70 (485)	25 (170)
F304N	80 (550)	35 (240)
F304LN	75 (515)	30 (205)
F309H	75 (515)	30 (205)
F310	75 (515)	30 (205)
F310H	75 (515)	30 (205)
F310MoLN	75 (515)	37 (225)
F316	75 (515)	30 (205)
F316H	75 (515)	30 (205)
F316L	70 (485)	25 (170)
F316N	80 (550)	35 (240)
F316LN	75 (515)	30 (205)
F317	75 (515)	30 (205)
F317L	70 (485)	25 (170)
F321	75 (515)	30 (205)
F321H	75 (515)	30 (205)
F347	75 (515)	30 (205)
F347H	75 (515)	30 (205)
F348	75 (515)	30 (205)
F348H	75 (515)	30 (205)
FXM-11	90 (620)	50 (345)
FXM-19	100 (690)	55 (380)
F10	80 (550)	30 (205)
F20	80 (550)	35 (240)
F44	94 (650)	44 (300)
F45	87 (600)	45 (310)
F46	78 (540)	35 (240)
F47	75 (515)	30 (205)
F48	80 (550)	35 (240)
F49	115 (795)	60 (415)
F56	73 (500)	27 (185)

(Continues)

(Continued)

Grade	Minimum Tensile Strength, ksi (MPa)	Minimum Yield Strength, ksi (MPa)
F58	109 (750)	61 (420)
F62	95 (655)	45 (310)
F50	100–130 (690–900)	65 (450)
F51	90 (620)	65 (450)
F52	100 (690)	70 (485)
F53	116 (800)	80 (550)
F54	116 (800)	80 (550)
F55	109–130 (750–895)	80 (550)
F57	118 (820)	85 (585)
F59	112 (770)	80 (550)
F60	95 (655)	70 (485)
F61	109 (750)	80 (550)

ASTM, A193/A193M-03 (Volume 01-01), Standard Specification for Alloy-Steel and Stainless Steel Bolting Materials for High-Temperature Service

Scope. This specification covers alloy and stainless steel bolting material for pressure vessels and flanges and fittings for high temperature service. The term *bolting material,* as used in the specification, covers bars, bolts, screws, studs, stud bolts, and wire. Bare and wire are hot wrought. The material may be further processed by centerless grinding or cold drawing. Austenitic stainless steel may be carbide treated and strain hardened.

Nuts for use with this bolting material are covered in ASTM A194/A194M.

Referenced Documents

ASTM

A194/A194M, Specification for Carbon and Alloy Steel Nuts for Bolts for High-Pressure or High-Temperature Service or Both.

A320/A320M, Specification for Alloy/Steel Bolting Materials for Low-Temperature Service.

A354, Specification for Quenched and Tempered Alloy Steel Bolts, Studs, and Other Externally Threaded Fasteners.

A962/A962M, Specification of Common Requirements for Steel Fasteners or Fastener Materials or Both, Intended for Use at Any Temperature from Cryogenic to the Creep Range.

E18, Test Methods for Rockwell Hardness and Rockwell Superficial Hardness of Metallic Materials.
E21, Test Methods for Elevated Temperature Tension Tests of Metallic Materials.
E112, Test Methods for Determining Average Grain Size.
E139, Test Methods for Conducting Creep, Creep-Rupture, and Stress-Rupture Tests of Metallic Materials.
E292, Test Methods for Conducting Time-for-Rupture Notch Tension Tests of Materials.
E328, Methods for Stress-Relaxation Tests for Materials and Structures.
E381, Method of Macroetch Testing Steel Bars, Billets, Blooms, and Forgings.
E566, Practice for Electromagnetic (Eddy-Current) Sorting of Ferrous Metals.
E709, Guide for Magnetic Particle Examination.
F606, Test Methods for Determining the Mechanical Properties of Externally and Internally Threaded Fasteners, Washers, and Rivets.
E150, Practice for Conducting Creep and Creep-Rupture Tension Tests of Metallic Materials under Conditions of Rapid Heating and Short Times.
E151, Practice for Tension Tests of Metallic Materials at Elevated Temperatures with Rapid Heating and Conventional or Rapid Strain Rates.

ASME

B1.1, Screw Threads.
B1.13M, Metric Screw Threads.
B18.2.1, Square and Hex Bolts and Screws.
B18.2.3.1M, Metric Hex Cap Screws.
B18.3, Hexagon Socket and Spline Socket Screws.
B18.3.1M, Metric Socket Head Cap Screws.

AIAG. AIAG B-5 02.00, Primary Metals Identification Tag Application Standard.

Sizes. Varies.

Methods of Manufacture. The steel shall be produced by any of the following processes: open hearth, basic oxygen, electric furnace, or vacuum induction melting (VIM).

Heat Treatment. Refer to ASTM A193/A193M.

Chemical Requirements (maximum in percentages). Refer to ASTM A193/A193M.

Mechanical Requirements. These are extracted from ASTM A193/A193M:

Grade	Minimum Tensile Strength, ksi (MPa)	Minimum Yield Strength, ksi (MPa)
B5		
Up to 4 in. (M100) inclusive	100 (690)	80 (550)
B6		
Up to 4 in. (M100) inclusive	110 (760)	85 (585)
B6X		
Up to 4 in. (M100) inclusive	90 (620)	70 (485)
B7		
$2\frac{1}{2}$ in. (M64) and under	125 (860)	105 (720)
$2\frac{1}{2}$ in. (M64) to 4 in. (100 mm)	115 (795)	95 (655)
Over 4 in. (M100) to 7 in. (175 mm)	100 (690)	75 (515)
B7M		
4 in. (M100) and under	100 (690)	80 (550)
Over 4 in. (M100) to 7 in. (M180)	100 (690)	75 (515)
B16		
$2\frac{1}{2}$ in. (M64) and under	125 (860)	105 (720)
$2\frac{1}{2}$ in. (M64) to 4 in. (M100)	110 (760)	95 (665)
Over 4 in. (M100) to 8 in. (M180)	100 (690)	85 (586)
Classes 1 and 1D: B8, B8M, B8P, B8LN, all diameters	75 (515)	30 (205)
Class 1: B8C, B8T, all diameters	75 (515)	30 (205)
Class 1A: B8A, B8CA, B8MA, B8PA, B8TA, B8LNA, B8MLNA, B8NA, B8MNA, B8NA, B8MNA, B8MLCuNA all diameters	75 (515)	30 (205)
Classes 1B and 1D: B8N, B8MN, B8MLCuN all diameters	80 (550)	35 (240)
Class 1C and 1D: B8R all diameters	100 (690)	55 (380)
Class 1C: B8RA all diameters	100 (690)	55 (380)
Class 1C and 1D: B8S all diameters	95 (655)	50 (345)
Class 1C: B8SA all diameters	95 (655)	50 (345)

Grade	Minimum Tensile Strength, ksi (MPa)	Minimum Yield Strength, ksi (MPa)
Class 2: B8, B8C, B8P, B8T, B8N[b]		
¾ in. and under	125 (860)	100 (690)
Over ¾ in. (M24) to 1 in. (M24)	115 (795)	80 (550)
Over 1 in. (M24) to 1¼ in. (M30)	105 (725)	65 (450)
Over 1¼ in. (M30) to 1½ in. (M36)	100 (690)	50 (345)
Class 2: B8M, B8MN, B8MLCuN[b]		
¾ in. (M20) and under	110 (760)	96 (665)
Over ¾ in. (M20) to 1 in. (M24)	100 (690)	80 (550)
Over 1 in. (M24) to 1¼ in. (M30)	95 (655)	65 (450)
Over 1¼ in. (M30) to 1½ in. (M36)	90 (620)	50 (345)
Class 2B: B8, B8M2 (note d)		
2 in. and under	95 (655)	75 (515)
Over 2 in. (M48) to 2½ in. (M64), inclusive	90 (620)	65 (450)
Over 2½ in. (M64) to 3 in.(M72), inclusive	80 (550)	55 (380)
Class 2C: B8M3[b] 2 in. (M48) and under	85 (585)	65 (450)
Over 2 in. (M48)	85 (585)	60 (415)

ASTM, A194/A194M-03b (Volume 01.01), Standard Specification for Carbon and Alloy Nuts for Bolts for High-Pressure or High-Temperature Service or Both

Scope. This specification covers a variety of carbon, alloy, and martensitic stainless steel nuts in the range ¼ in. through 4 in. (metric M6 through M100 nominal). It also covers austenitic steel nuts in the size range ¼ in. (M6 nominal) and above. These nuts are intended for high-pressure or high-temperature service or both.

Referenced Documents

ASTM

A153/A153M, Specification for Zinc Coating (Hot-Dip) on Iron and Steel Hardware.

A276, Specification for Stainless Steel Bars and Shapes.

A320/A320M, Specification for Alloy-Steel Bolting Materials for Low-Temperature Service.
A370, Test Methods and Definitions for Mechanical Testing of Steel Products.
A962/A962M, Specification of Common Requirements for Steel Fasteners or Fastener Materials or Both, Intended for Use at Any Temperature from Cryogenic to the Creep Range.
B633, Specification for Electrodeposited Coatings of Zinc on Iron and Steel.
B695, Specification for Coatings of Zinc Mechanically Deposited on Iron and Steel.
B696, Specification for Coatings of Cadmium Mechanically Deposited.
B766, Specification for Electrodeposited Coatings of Cadmium.
E112, Test Methods for Determining Average Grain Size.

ASME

B1.1, Unified Screw Threads.
B1.13M, Metric Screw Threads.
B18.2.2, Square and Hex Nuts.
B18.2.4.6M, Metric Heavy Hex Nuts.

Methods of Manufacture. From hot wrought bars.

Heat Treatment. Refer to ASTM A194/A194M.

Chemical Requirements (maximums). Refer to ASTM A 194/A19M.

Mechanical Requirements. Refer to ASTM A194/A19M.

Hardness Requirements. These are extracted from ASTM A 194/A 194M:

Grade	Brinell
1	121 minimum
2	159–352
2H	
≤1½ in. (M36)	248–352
Over 1½ in. (M36)	212–352
2HM, 7M	159–237
3, 4, 7, and 16	248–352
6 and 6F	228–271
8, 8C, 8M, 8T, 8F, 8P, 8N	126–300
8MN, 8LN, 8MLN, 8MLCuN, and 9C	126–300

Grade	Brinell
8A, 8CA, 8MA, 8TA	126–192
8FA, 8PA, 8NA, 8MNA	126–192
8LNA, 8MLNA, & 8MLCuNA	126–192
8R, 8RA, 8S, and 8SA	183–271
9C, 9CA	126–192

ASTM, A202/A202M-03 (Volume 01.04), Standard Specification for Pressure Vessel Plates, Alloy Steel, Chromium-Manganese-Silicon

Scope. This specification covers chromium-manganese-silicon alloy steel plates, intended particularly for welded boilers and other pressure vessels. Plates under this specification are available in two grades having strength levels as follows:

Grade	Tensile Strength, ksi (MPa)
A	75–95 (515–655)
B	85–110 (585–760)

The maximum thickness of plates is limited only by the capacity of the composition to meet the specified mechanical property requirements; however, current practice normally limits the maximum thickness of plates furnished under this specification to 2 in. (50 mm).

Grade A is suitable for rivets, and when so used, the bars are subject to the requirements for rolled bars specified in Specification A31, except for the chemical and mechanical requirements.

Referenced Documents

ASTM

A31, Specification for Steel Rivets and Bars for Rivets, Pressure Vessels.
A20/A20M, Specification for General Requirements for Steel Plates for Pressure Vessels.
A435/A435M, Specification for Straight-Beam Ultrasonic Examination of Steel Plates.
A577/A577M, Specification for Ultrasonic Angle-Beam Examination of Steel Plates.

A578/A578M, Specification for Straight-Beam Ultrasonic Examination of Plain and Clad Steel Plates for Special Applications.

Methods of Manufacture. The steel shall be killed.

Heat Treatment. The plates may be supplied normalized or stress relieved or both.

Chemical Requirements. Refer to ASTM A202/A202M.

Mechanical Requirements. These are extracted from ASTM A202/A202M:

Material	Grade	Tensile Strength, ksi (MPa)	Minimum Yield Strength, ksi (MPa)
A 202	A	75–95 (515–655)	45 (310)
	B	85–110 (585–760)	47 (325)

ASTM, A203/A203M-97(2003) (Volume 01.04), Standard Specification for Pressure Vessel Plates, Alloy Steel, Nickel

Scope. This specification covers nickel-alloy steel plates intended primarily for welded pressure vessels. Plates under this specification are available with four strength levels and two nickel compositions as follows:

Grade	Nominal Nickel Content, %	Minimum Yield Strength, ksi (MPa)	Minimum Tensile Strength, ksi (MPa)
A	2.25	37 (255)	65 (450)
B	2.25	40 (275)	70 (485)
D	3.50	37 (255)	65 (450)
E	3.50	40 (275)	70 (485)
F	3.50		
2 in. (50 mm) and under		55 (380)	80 (550)
Over 2 in. (50 mm)		50 (345)	75 (515)

The maximum thickness of plates is limited only by the capacity of the composition to meet the specified mechanical property requirements.

However, current practice normally limits the maximum thickness of plates furnished under this specification as follows:

Grade	Maximum Thickness, in. (mm)
A	6 (150)
B	6 (150)
D	4 (100)
E	4 (100)
F	4 (100)

Referenced Documents

ASTM

A20/A20M, Specification for General Requirements for Steel Plates for Pressure Vessels.
A435/A435M, Specification for Straight-Beam Ultrasonic Examination of Steel Plates.
A577/A577M, Specification for Ultrasonic Angle-Beam Examination of Steel Plates.
A578/A578M, Specification for Straight-Beam Ultrasonic Examination of Plain and Clad Steel Plates for Special Applications.

Methods of Manufacture. The steel is killed and conforms to the fine grain size requirement of Specification AS.

Heat Treatment. Refer to ASTM A203/A203M.

Chemical Requirements (maximums). Refer to ASTM 203/A203M.

Mechanical Requirements. These are extracted from ASTM A202/A202M:

Material	Grade	Tensile Strength, ksi (MPa)	Minimum Yield Strength, ksi (MPa)
A 203	A and D	≤2 in.: 65–85 (450–585)	≤2 in.: 37 (255)
		Over 2 in.: 65–85 (450–585)	Over 2 in.: 37 (255)
	B and E	≤2 in.: 70–90 (485–620)	≤2 in.: 40 (275)
		Over 2 in.: 70–90 (485–620)	Over 2 in.: 40 (275)
	F	≤2 in.: 80–100 (550–690)	≤2 in.: 55 (380)
		Over 2 in.: 75–95 (515–655)	Over 2 in.: 50 (345)

ASTM, A204/A204M-03 (Volume 01.04), Standard Specification for Pressure Vessel Plates, Alloy Steel, Molybdenum

Scope. This specification covers molybdenum-alloy steel plates, intended particularly for welded boilers and other pressure vessels. Plates under this specification are available in three grades having different strength levels as follows:

Grade	Tensile Strength, ksi (MPa)
A	65–85 (450–585)
B	70–90 (485–620)
C	75–95 (515–655)

The maximum thickness of plates is limited only by the capacity of the composition to meet the specified mechanical property requirements; however, current practice normally limits the maximum thickness of plates furnished under this specification as follows:

Grade	Maximum Thickness, in. (mm)
A	6 (150)
B	6 (150)
C	4 (100)

Referenced Documents

ASTM

A20/A20M, Specification for General Requirements for Steel Plates for Pressure Vessels.
A435/A435M, Specification for Straight-Beam Ultrasonic Examination of Steel Plates.
A577/A577M, Specification for Ultrasonic Angle-Beam Examination of Steel Plates.
A578/A578M, Specification for Straight-Beam Ultrasonic Examination of Plain and Clad Steel Plates for Special Applications.

Sizes. The maximum thickness of plates is limited only by the capacity of the composition to meet the specified mechanical property requirements.

Methods of Manufacture. The steel is killed.

Heat Treatment. Plates $1\frac{1}{2}$ in. (40 mm) and under in thickness are normally supplied in the as-rolled condition. The plates may be ordered normalized or stress relieved. Plates over $1\frac{1}{2}$ in. (40 mm) in thickness are normalized.

Chemical Requirements. Refer to ASTM A204/A204M.

Mechanical Requirements. These are extracted from ASTM A204/A204M:

Material	Grade	Tensile Strength, ksi (MPa)	Yield Strength, ksi (MPa)
A 204	A	65–85 (450–585)	37 (255)
	B	70–90 (485–620)	40 (275)
	C	75–95 (515–655)	43 (295)

ASTM, A216/A216M-93 (2003) (Volume 01.02), Standard Specification for Steel Castings, Carbon, Suitable for Fusion Welding, for High-Temperature Service

Scope. This specification covers carbon steel castings for valves, flanges, fittings, or other pressure-containing parts for high-temperature service and of quality suitable for assembly with other castings or wrought-steel parts by fusion welding.

Three grades, WCA, WCB, and WCC, are covered in this specification. Selection depends on the design and service conditions, mechanical properties, and high temperature characteristics.

Referenced Documents

ASTM

A488/A488M, Practice for Steel Castings, Welding, Qualifications of Procedures and Personnel.
A703/A703M, Specification for Steel Castings, General Requirements, for Pressure-Containing Parts.
E165, Test Method for Liquid Penetrant Examination.
E709, Guide for Magnetic Particle Examination.

MSS (Manufacturer's Standardization Society). SP-55 Steel Castings for Valve, Flanges, and Fittings, and Other Components (Visual Method).

Sizes. Varies.

Methods of Manufacture. The steel is made by the electric furnace process with or without separate refining such as argon-oxygen decarburization (AOD).

Heat Treatment. Refer to ASTM A216/A216M.

Welding Repair. Repair procedures and welder qualifications are to ASTM A488/A488M.

Chemical Requirements. Refer to ASTM A216/A216M.

Mechanical Requirements. These are extracted from ASTM A216/A216M:

Grade	Minimum Tensile Strength, ksi (MPa)	Minimum Yield Strength, ksi (MPa)
WCA	60–85 (415–585)	30 (205)
WCB	70–95 (485–655)	36 (250)
WCC	70–95 (485–655)	40 (275)

ASTM, A217/A217M-02 (Volume 01.02), Standard Specification for Steel Castings, Martensitic Stainless and Alloy, for Pressure-Containing Parts, Suitable for High-Temperature Service

Scope. This specification covers martensitic stainless steel and alloy steel castings for valves, flanges, fittings, and other pressure-containing parts (Note: Carbon steel castings for pressure-containing parts are covered by Specification A216/A216M. Low-alloy quench-and-tempered grades equivalent to Specification A217/A217M grades may be found in both Specifications A352/A352M and A487/A487M) intended primarily for high-temperature and corrosive service.

One grade of martensitic stainless steel and nine grades of ferritic alloy steel are covered. Selection depends on the design and service conditions, mechanical properties, and the high-temperature and corrosion-resistant characteristics.

Referenced Documents

ASTM

A216/A216M, Specification for Steel Castings, Carbon, Suitable for Fusion Welding, for High-Temperature Service.
A352/A352M, Specification for Steel Castings, Ferritic and Martensitic, for Pressure-Containing Parts, Suitable for Low-Temperature Service.
A487/A487M, Specification for Steel Castings Suitable for Pressure Service.
A488/A488M, Practice for Steel Castings, Welding, Qualifications of Procedures and Personnel.
A703/A703M, Specification for Steel Castings, General Requirements, for Pressure-Containing Parts.
A802/A802M, Practice for Steel Castings, Surface Acceptance Standards, Visual Examination.
E165, Test Method for Liquid Penetrant Examination.
E709, Guide for Magnetic Particle Examination.

Sizes. Varies.

Methods of Manufacture. The steel is made by the electric furnace process, with or without separate refining, such as argon-oxygen decarburization.

Heat Treatment. Refer to ASTM A217/A217M.

Welding Repair. Repair procedures and welder qualifications are in ASTM A488/A488M.

Chemical Requirements. Refer to extract from ASTM A217/A217M.

Mechanical Requirements. These are extracted from ASTM A217/A217M:

Grade	Minimum Tensile Strength, ksi (MPa)	Minimum Yield Strength, ksi (MPa)
WC1	65–90 (450–620)	35 (240)
WC4, WC5, WC6, WC9	70–95 (485–655)	40 (275)
WC11	80–105 (550–745)	50 (345)
C5, C12	90–115 (620–760)	60 (415)
C12A	85–110 (585–760)	60 (415)
CA15	90–115 (620–795)	60 (415)

ASTM, A234/A234M-03 (Volume 01.01), Standard Specification for Piping Fittings of Wrought Carbon Steel and Alloy Steel for Moderate- and High-Temperature Service

Scope. This specification covers wrought carbon steel and alloy steel fittings of seamless and welded construction covered by the latest revision of ASME B16.9, B16.11, B16.28, MSS SP-79, and MSS SP-95. These fittings are for use in pressure piping and pressure vessel fabrication for service at moderate and elevated temperatures.

Referenced Documents

ASTM

A216/A216M, Specification for Steel Castings, Carbon, Suitable for Fusion Welding, for High-Temperature Service.

A217/A217M, Specification for Steel Castings, Martensitic Stainless and Alloy, for Pressure-Containing Parts Suitable for High-Temperature Service.

A960, Specification for Common Requirements for Wrought Steel Piping Fittings.

ASME

B16.9, Steel Butt-Welding Fittings.
B16.11, Forged Steel Fittings, Socket Welding and Threaded.
B16.28, Wrought Steel Butt-Welding Short Radius Elbows and Returns.

ASME BPV Code

Section V, Nondestructive Examination.
Section VIII, Division 1, Pressure Vessels.
Section IX, Welding Qualifications.

MSS

SP-25, Standard Marking System for Valves, Fittings, Flanges, and Unions.
SP-79, Socket Welding Reducer Inserts.
SP-95, Swage(d) Nipples and Bull Plugs.

ANSI. SNT-TC-1A (1984), Recommended Practice for Nondestructive Testing Personnel Qualification and Certification.

Methods of Manufacture. Shaping operations are performed by hammering, pressing, piercing upsetting, rolling bending, fusion welding, machining, or by a combination of two or more of these operations.

Heat Treatment. Refer to ASTM A234/A234M.

Welding Repair. Repair procedures and welder qualifications are in ASTM A488/A488M.

Chemical Requirements. Refer to ASTM A234/A234M.

Mechanical Requirements. These are extracted from ASTM A234/A234M:

Grade	Minimum Tensile Strength, ksi (MPa)	Minimum Yield Strength, ksi (MPa)
WPB	60–85 (415–585)	35 (240)
WPC, WP11 Cl 2, WP12 Cl 2	70–95 (485–655)	40 (275)
WP1	55–80 (380–550)	30 (205)
WP11 Cl 1, WP22 Cl 1, WP5 Cl 1, WP9 Cl 1	60–85 (415–585)	30 (205)
WPR	63–88 (435–605)	46 (315)
WP11 Cl 3, WP22 Cl 3, WP5 Cl 3, WP9 Cl 3	75–100 (520–690)	45 (310)
WP91	85–110 (585–760)	60 (415)
WP911	90–120 (620–840)	64 (440)
WP12 Cl 1	60–85 (415–585)	32 (220)

ASTM, A285/A285M-03 (Volume 01.04), Standard Specification for Pressure Vessel Plates, Carbon Steel, Low- and Intermediate-Tensile Strength

Scope. This specification covers carbon steel plates of low- and intermediate-tensile strengths, which may be made by killed, semi-killed, capped, or rimmed steel practices, at the producer's option. These plates are intended for fusion-welded pressure vessels. Plates under this specification are available in three grades, having different strength levels as follows:

Grade	Tensile Strength, ksi (MPa)
A	45–65 (310–450)
B	50–70 (345–485)
C	55–75 (380–515)

Referenced Document

ASTM. A20/A20M, Specification for General Requirements for Steel Plates for Pressure Vessels.

Methods of Manufacture Refer to ASTM A285/A285M.

Heat Treatment. The plates are supplied normalized, stress relieved, or both.

Chemical Requirements. Refer to ASTM A285/A285M.

Mechanical Requirements. These are extracted from ASTM A285/A285M:

Material	Grade	Tensile Strength, ksi (MPa)	Minimum Yield Strength, ksi (MPa)
A 285	A	45–65 (310–450)	24 (165)
	B	50–70 (345–485)	27 (185)
	C	55–75 (380–515)	30 (205)

ASTM, A302/A302M-03 (Volume 01.04), Standard Specification for Pressure Vessel Plates, Alloy Steel, Manganese-Molybdenum, and Manganese-Molybdenum-Nickel

Scope. This specification covers manganese-molybdenum and manganese-molybdenum-nickel alloy steel plates intended particularly for welded boilers and other pressure vessels. Plates under this specification are available in four grades having different strength levels as follows:

Grade	Tensile Strength, ksi (MPa)	Type
A	75–95 (515–655)	Mn-Mo
B	80–100 (550–690)	Mn-Mo
C	80–100 (550–690)	Mn-Mo-Ni
D	80–100 (550–690)	Mn-Mo-Ni

The maximum thickness of plates is limited only by the capacity of the chemical composition to meet the specified mechanical property requirements. The minimum thickness is limited to 0.25 in. (6.5 mm).

Referenced Document

ASTM. A20/A20M, Specification for General Requirements for Steel Plates for Pressure Vessels.

Chemical Composition. Refer to ASTM A302/A302M.

Mechanical Requirements. These are extracted from ASTM A302/A302M:

Material	Grade	Tensile Strength, ksi (MPa)	Minimum Yield Strength, ksi (MPa)
A 302	A	75–95 (515–655)	45 (310)
	B	80–100 (550–690)	50 (345)
	C	80–100 (550–690)	50 (345)
	D	80–100 (550–690)	50 (345)

ASTM, A307-03 (Volume 01.08), Standard Specification for Carbon Steel Bolts and Studs, 60,000 psi Tensile Strength

Scope. This specification covers the chemical and mechanical requirements of three grades of carbon steel bolts and studs in sizes ¼ in. (6.35 mm) through 4 in. (104 mm). The fasteners are designated by "Grade," denoting tensile strength and intended use, as follows:

Grade	Description
Grade A	Bolts and studs having a minimum tensile strength of 60 ksi (414 MPa), intended for general applications
Grade B	Bolts and studs having a tensile strength of 60–100 ksi (414–690 MPa), intended for flanged joints in piping systems with cast iron flanges
Grade C	Nonheaded anchor bolts, either bent or straight, having properties conforming to Specification A36 (tensile strength of 58–80 ksi (400–550 MPa), intended for structural anchorage purposes

The term *studs* includes stud stock, sometimes referred to as *threaded rod*.

Suitable nuts are covered in Specification A563. Unless otherwise specified, the grade and style of nut for each grade of fastener, of all surface finishes, is as follows:

Fastener Grade and Size	Nut Grade and Style
A, C, ¼–1½ in.	A, hex
A, C, over 1½–4 in.	A, heavy hex
B, ¼–4 in.	A, heavy hex nuts of other grades and styles having specified proof load stresses

Referenced Documents

ASTM

A370, Test Methods and Definitions for Mechanical Testing of Steel Products.
A563, Specification for Carbon and Alloy Steel Nuts.
A706/A706M, Specification for Low-Alloy-Steel Deformed Bars for Concrete Reinforcement.
A751, Test Methods, Practices, and Terminology for Chemical Analysis of Steel Products.
B695, Specification for Coatings of Zinc Mechanically Deposited on Iron and Steel.
D3951, Practice for Commercial Packaging.
F606, Test Methods for Determining the Mechanical Properties of Externally and Internally Threaded Fasteners, Washers, and Rivets.
F1470, Guide for Fastener Sampling for Specified Mechanical Properties and Performance Inspection.
A36/A36M, Specification for Carbon Structural Steel.
A153, Specification for Zinc Coating (Hot-Dip) on Iron and Steel Hardware.

ASME

B1.1, Unified Screw Threads.
B18.2.1, Square and Hex Bolts and Screws.
B18.24.1, Part Identifying Number (PIN) Code System.
MIL-STD105, Single Sampling Plan for Normal Inspection.

Methods of Manufacture. Open hearth, basic oxygen, or electric furnace process.

Heat Treatment. Refer to ASTM A307/A307M.

Threads. Unless specified otherwise, threads are in the coarse thread series, as specified in the latest issue of ASME B1.1 and have a Class 2A tolerance.

Chemical Requirements. Refer to ASTM A307/A307M.

Mechanical Requirements. These are extracted from ASTM A307/A307M:

Material	Grade	Minimum Tensile Strength, ksi (MPa)	Minimum Yield Strength, ksi (MPa)
A 307	A	60 (415)	
	B	60–100 (415–690)	
	C	58–80 (400–550)	36 (50)

ASTM, A312/A312M-03 (Volume 01.01), Standard Specification for Seamless and Welded Austenitic Stainless Steel Pipes

Scope. This specification covers seamless, straight-seam welded, and heavily cold-worked-welded austenitic stainless steel pipe intended for high-temperature and general-corrosive service.

Grades TP304H, TP309H, TP309HCb, TP310H, TP310HCb, TP316H, TP321H, TP347H, and TP348H are modifications of Grades TP304, TP309Cb, TP309S, TP310Cb, TP310S, TP316, TP321, TP347, and TP348, intended for high-temperature service.

Referenced Documents

ASTM

A262, Practices for Detecting Susceptibility to Intergranular Attack in Austenitic Stainless Steels.
A370, Test Methods and Definitions for Mechanical Testing of Steel Products.
A941, Terminology Relating to Steel, Stainless Steel, Related Alloys, and Ferroalloys.
A999/A999M, Specification for General Requirements for Alloy and Stainless Steel Pipe.
A1016/A1016M, Specification for General Requirements for Ferritic Alloy Steel, Austenitic Alloy Steel, and Stainless Steel Tubes.
E112, Test Methods for Determining the Average Grain Size.
E381, Method of Macroetch Testing Steel Bars, Billets, Blooms, and Forgings.
E527, Practice for Numbering Metals and Alloys (UNS).

ASME

B1.20.1, Pipe Threads, General Purpose.
B36.10, Welded and Seamless Wrought Steel Pipe.
B36.19, Stainless Steel Pipe.

ASME BPV Code. Section VIII. Unfired Pressure Vessels

AWS. A5.9, Corrosion-Resisting Chromium and Chromium-Nickel Steel Welding Rods and Electrodes.

Society for Automative Engineers (SAE). J1086, Practice for Numbering Metals and Alloys (UNS).

ANSI. SNT-TC-1A, Personnel Qualification and Certification in Nondestructive Testing.

Sizes. Varies.

Methods of Manufacture. The pipe is manufactured by one of the following processes:

Seamless (SMLS) *pipe* is made by a process that does not involve welding at any stage of production.

Welded (WLD) *pipe* is made using an automatic welding process with no addition of filler during the welding process.

Heavily cold-worked (HCW) *pipe* is made by applying cold working of not less than 35% reduction in thickness of both wall and weld to a welded pipe prior to the final annealing. No filler is used in making the weld.

Heat Treatment. Refer to ASTM A312/A312M.

Chemical Requirements. Refer to ASTM A312/A312M.

Mechanical Requirements. These are extracted from ASTM A312/A312M:

Grade	Minimum Tensile Strength, ksi (MPa)	Minimum Yield Strength, ksi (MPa)
TP304L	70 (485)	25 (170)
TP316L	70 (485)	25 (170)
TP304	75 (515)	30 (205)
TP304H	75 (515)	30 (205)
TP309Cb	75 (515)	30 (205)

(Continues)

(Continued)

Grade	Minimum Tensile Strength, ksi (MPa)	Minimum Yield Strength, ksi (MPa)
TP309H	75 (515)	30 (205)
TP309HCb	75 (515)	30 (205)
TP309S	75 (515)	30 (205)
TP310Cb	75 (515)	30 (205)
TP310H	75 (515)	37 (225)
TP310Cb	75 (515)	37 (225)
TP310S	75 (515)	30 (205)
TP316	75 (515)	30 (205)
TP316H	75 (515)	30 (205)
TP317	75 (515)	30 (205)
TP317L	75 (515)	30 (205)
TP321		
Welded	75 (515)	30 (205)
Seamless		
≤3/8 in.	75 (515)	30 (205)
>3/8 in.	70 (485)	25 (170)
TP321H		
Welded	75 (515)	30 (205)
Seamless		
≤3/16 in.	75 (515)	30 (205)
>3/16 in.	70 (480)	25 (170)
TP347	75 (515)	30 (205)
TP347H	75 (515)	30 (205)
TP348	75 (515)	30 (205)
TP348H	75 (515)	30 (205)
TPXM-10	90 (620)	50 (345)
TPXM-11	90 (620)	50 (345)
TPXM-15	75 (515)	30 (205)
TPXM-29	100 (690)	55 (380)
TPXM-19	100 (690)	55 (380)
TP304N	80 (550)	35 (240)
TP316N	80 (550)	35 (240)
TP304LN	75 (515)	30 (205)
TP316LN	75 (515)	30 (205)

ASTM, A320/A320M-03 (Volume 01.01), Standard Specification for Alloy-Steel Bolting Materials for Low-Temperature Service

Scope. This specification covers alloy steel bolting materials for pressure vessels, valves, flanges, and fittings for low-temperature service. The term *bolting material*, as used in this specification, covers rolled, forged, or strain-hardened bars, bolts, screws, studs, and stud bolts. The bars are hot-wrought. The material may be further processed by centerless grinding or cold drawing. Austenitic stainless steel is solution annealed or annealed and strain hardened.

Several grades are covered, including both ferritic and austenitic steels designated L7, B8, etc. Selection depends on the design, service conditions, mechanical properties, and low-temperature characteristics. The mechanical requirements of the following table indicate the diameters for which the minimum mechanical properties apply to the various grades and classes, and Table 2 (in the specification) stipulates the requirements for Charpy impact energy absorption. The manufacturer should determine that the material can conform to these requirements before parts are manufactured. For example, when Grade L43 is specified to meet the Table 2 impact energy values at −150°F (−101°C), additional restrictions (such as procuring a steel with lower P and S contents than might normally be supplied) in the chemical composition for AISI 4340 are likely to be required.

Nuts for use with this bolting material are covered in ASTM A194/A194M and the nut material should be impact tested.

Referenced Documents

ASTM

A194/A194M, Specification for Carbon and Alloy Steel Nuts for Bolts for High-Pressure or High-Temperature Service or Both.
A370, Test Methods and Definitions for Mechanical Testing of Steel Products.
A962/A962M, Specification of Common Requirements for Steel Fasteners or Fastener Materials or Both, Intended for Use at Any Temperature from Cryogenic to the Creep Range.
E566, Practice for Electromagnetic (Eddy-Current) Sorting of Ferrous Metals.
F436, Specification for Hardened Steel Washers.

ASME

B1.1, Screw Threads.
B18.22.1, Plain Washers.

Sizes. Varies.

Methods of Manufacture. Refer to ASTM A320/A320M.

Heat Treatment. Refer to ASTM A320/A320M.

Chemical Requirements. Refer to ASTM A320/A320M.

Mechanical Requirements. These are extracted from ASTM A320/A320M:

Grade	Minimum Tensile Strength, ksi (MPa)	Minimum Yield Strength, ksi (MPa)
Ferritic Steels:		
L7, L7A, L7B, L7C, L70, L71, L72, L73	125 (860)	105 (725)
2 1/2 in. (65 mm) and under L43		
4 in. (100 mm) and under L7M	125 (860)	105 (725)
2 1/2 in. (65 mm) and under L1	100 (690)	80 (550)
1 in. (25 mm) and under	125 (860)	105 (725)
Austenitic Steels:		
Class 1: B8, B8C, B8M, B8P, B8F, B8T, B8LN, B8MLN: all diameters	75 (515)	30 (205)
Class 1A: B8A, B8CA, B8MA, B8PA, B8FA, B8TA, B8LNA, B8MLNA: all diameters	75 (515)	30 (205)
Class 2: B8, B8C, B8P, B8T:		
3/4 in. (20 mm) and under	125 (860)	100 (690)
Over 3/4–1 in. (20–25 mm) inclusive.	115 (795)	80 (550)
Over 1 in. to 1 1/4 in (25–32 mm) inclusive	105 (725)	65 (450)
Over 1 1/4 in. to 1 1/2 in. (32–40 mm) inclusive	100 (690)	50 (345)
Class 2: B8M:		
3/4 in. (20 mm) and under	110 (760)	95 (655)
Over 3/4 in. to 1 in. (20–25 mm) inclusive	100 (690)	80 (550)
Over 1 in. to 1 1/4 in (25–32 mm) inclusive	95 (655)	65 (450)
Over 1 1/4 in. to 1 1/2 in. (32 mm to 40 mm) inclusive	90 (620)	50 (345)

ASTM, A333/A333M-99 (Volume 01.01), Standard Specification for Seamless and Welded Steel Pipe for Low-Temperature Service

Scope. This specification covers nominal (average) wall seamless and welded carbon and alloy steel pipe intended for use at low temperatures.

Several grades of ferritic steel are included as listed in the following table. Some product sizes may not be available under this specification because heavier wall thicknesses have an adverse affect on low-temperature impact properties.

Referenced Documents

ASTM

A370, Test Methods and Definitions for Mechanical Testing of Steel Products.
A530/A530M, Specification of General Requirements for Specialized Carbon and Alloy Steel Pipe.
A671, Specification for Electric-Fusion-Welded Steel Pipe for Atmospheric and Lower Temperatures.
E23, Test Methods for Notched Bar Impact Testing of Metallic Materials.
E213, Practice for Ultrasonic Examination of Metal Pipe and Tubing.
E309, Practice for Eddy-Current Examination of Steel Tubular Products Using Magnetic Saturation.

Sizes. Nominal pipe size ⅛–48 in.

Methods of Manufacture. Refer to ASTM A333/A333M.

Chemical Requirements. Refer to ASTM A333/A333M.

Mechanical Requirements. These are extracted from ASTM A333/A333M:

Grade	Minimum Tensile Strength, ksi (MPa)	Minimum Yield Strength, ksi (MPa)
1	55.0 (380)	30.0 (205)
3	65.0 (450)	35.0 (240)
4	60.0 (415)	35.0 (240)
6	60.0 (415)	35.0 (240)
7	60.0 (415)	35.0 (240)
8	100.0 (690)	75.0 (515)
9	63.0 (435)	46.0 (315)
10	80.0 (550)	65.0 (450)
11	65.0 (450)	35.0 (240)

ASTM, A335/A335M-03 (Volume 01.01), Standard Specification for Seamless Ferritic Alloy-Steel Pipe for High-Temperature Service

Scope. This specification covers nominal (average) wall seamless alloy-steel pipe intended for high-temperature service. Pipe ordered to this specification should be suitable for bending, flanging (vanstoning) and similar forming operations, and fusion welding. Selection depends on the design, service conditions, mechanical properties, and high-temperature characteristics.

Several grades of ferritic steels (Note: Ferritic steels in this specification are defined as low- and intermediate-alloy steels containing up to and including 10% chromium) are covered. Their compositions are given in the following table.

Referenced Documents

ASTM

A450/A450M, Specification of General Requirements for Carbon, Ferritic Alloy, and Austenitic Alloy Steel Tubes.
A999/A999M, Specification of General Requirements for Alloy and Stainless Steel Pipe.
E213, Practice for Ultrasonic Examination of Metal Pipe and Tubing.
E309, Practice for Eddy-Current Examination of Steel Tubular Products Using Magnetic Saturation.
E381, Method of Macroetch Testing Steel Bars, Billets, Blooms, and Forgings.
E527, Practice for Numbering Metals and Alloys (UNS).
E570, Practice for Flux Leakage Examination of Ferromagnetic Steel Tubular Products.

ANSI. SNT-TC-1A, Recommended Practice for Nondestructive Testing Personnel Qualification and Certification.

SAE. J1086, Practice for Numbering Metals and Alloys (UNS).

Sizes. Nominal pipe size $^1/_8$–48 in.

Chemical Requirements. Refer to ASTM A335/A335M.

Mechanical Requirements. These are extracted from ASTM A335/A335M:

Grade	Minimum Tensile Strength, ksi (MPa)	Minimum Yield Strength, ksi (MPa)
P1, P2	55.0 (380)	30.0 (205)
P12	60.0 (415)	32.0 (220)
P23	74.0 (510)	58.0 (400)
P91	85.0 (585)	60.0 (415)
P92, P911	90.0 (620)	64.0 (440)
P122	90.0 (620)	58.0 (400)
All others	60.0 (415)	30.0 (205)

ASTM, A350/A350M-02b (Volume 01.01), Standard Specification for Carbon and Low-Alloy Steel Forgings, Requiring Notch Toughness Testing for Piping Components

Scope. This specification covers several grades of carbon and low-alloy steel forged or ring-rolled flanges, forged fittings, and valves intended primarily for low-temperature service and requiring notch toughness testing. They are made to specified dimensions, or dimensional standards, such as the ASME and API Specifications referenced next. Although this specification covers some piping components machined from rolled bar and seamless tubular materials, it does not cover raw material produced in these product forms.

No limitation on size is intended beyond the ability of the manufacturer to obtain the specified requirements. However, Class 3 of Grade LF787 is available only in the quenched-and-precipitation heat-treated condition.

Referenced Documents

ASTM

A370, Test Methods and Definitions for Mechanical Testing of Steel Products.
A788, Specification for Steel Forgings, General Requirements.
A961, Specification for Common Requirements for Steel Flanges, Forged Fittings, Valves, and Parts for Piping Applications.

ASME

B16.5, Steel Pipe Flanges and Flanged Fittings.

B16.9, Factory-made Wrought Steel Butt-Welding Fittings.
B16.10, Face-to-Face and End-to-End Dimensions of Ferrous Valves.
B16.11, Forged Steel Fittings, Socket-Welding and Threaded.
B16.30, Unfired Pressure Vessel Flange Dimensions.
B16.34, Valves-Flanged, Threaded, and Welding End.
B16.47, Large Diameter Steel Flanges.

ASME BPV. Section IX. Welding Qualifications.

(AWS).

A5.1, Mild Steel Covered Arc-Welding Electrodes.
A5.5, Low-Alloy Steel Covered Arc-Welding Electrodes.

API

600, Steel Gate Valves with Flanged or Butt-Welding Ends.
602, Compact Design Carbon Steel Gate Valves for Refinery Use.
605, Large Diameter Carbon Steel Flanges.

Methods of Manufacture. The steel is made by any of the following primary processes: open hearth, basic oxygen, electric furnace or vacuum-induced melting.

Heat Treatment. Refer to ASTM A350/A350M.

Welding Repair. Repair procedures and welder qualifications are in ASME Section IX of the code.

Hardness. Except when only one forging is produced, a minimum of two forgings are hardness tested per batch or continuous run to ensure that the hardness of the forgings does not exceed 197 HB after heat treatment for the mechanical properties.

Chemical Requirements. Refer to ASTM A350/A350M.

Mechanical Requirements. These are extracted from ASTM A350/ A350M:

Grade	Minimum Tensile Strength, ksi (MPa)	Minimum Yield Strength, ksi (MPa)
LF1, LF5 Class 1	60–85 (415–585)	30 (205)
LF2 Classes 1 and 2	70–95 (485–655)	36 (250)
LF3 and LF5 Class 2	70–95 (485–655)	37.5 (260)
LF6 Class 1	66–91 (455–630)	52 (360)
LF6 Classes 2 and 3	75–100 (515–690)	60 (415)
LF9	63–88 (435–605)	46 (315)
LF787 Class 2	65–85 (450–585)	55 (380)
LF787 Class 3	75–95 (515–655)	65 (450)

ASTM, A351/A351M-03 (Volume 01.02), Standard Specification for Castings, Austenitic, Austenitic-Ferritic (Duplex), for Pressure-Containing Parts

Scope. This specification covers austenitic and austenitic-ferritic (duplex) steel castings for valves, flanges, fittings, and other pressure-containing parts (Note: Carbon steel castings for pressure-containing parts are covered by Specification A 216/A216M and low-alloy steel castings by Specification A 217/A217M).

A number of grades of austenitic and austenitic-ferritic steel castings are included in this specification. Since these grades possess varying degrees of suitability for service at high temperatures or in corrosive environments, it is the responsibility of the purchaser to determine which grade is furnished. Selection depends on the design and service conditions, mechanical properties, and high-temperature or corrosion-resistant characteristics or both.

Because of thermal instability, Grades CE20N, CF3A, CF3MA, and CF8A are not recommended for service at temperatures above 800°F (425°C).

Because of embrittlement phases, Grade CD4MCu is not recommended for service at temperatures above 600°F (316°C).

Referenced Documents

ASTM

A216/A216M, Specification for Steel Castings, Carbon, Suitable for Fusion Welding, for High-Temperature Service.

A217/A217M, Specification for Steel Castings, Martensitic Stainless and Alloy, for Pressure-Containing Parts, Suitable for High-Temperature Service.
A488/A488M, Practice for Steel Castings, Welding, Qualification of Procedures and Personnel.
A703/A703M, Specification for Steel Castings, General Requirements, for Pressure-Containing Parts.
E165, Test Method for Liquid Penetrant Examination.
E709, Guide for Magnetic Particle Examination.

MSS. SP-55, Quality Standard for Steel Castings for Valves, Flanges, and Fittings and Other Components (Visual Method).

Sizes. Varies.

Methods of Manufacture. The steel is made by the electric furnace process with or without separate refining such as argon-oxygen decarburization.

Heat Treatment. Refer to ASTM A351/A351M.

Welding Repair. Repair procedures and welder qualifications shall be to ASTM A488/A488M.

Chemical Requirements. Refer to ASTM A351/A351M.

Mechanical Requirements. These are extracted from ASTM A351/A351M:

Grade	Minimum Tensile Strength, ksi (MPa)	Minimum Yield Strength, ksi (MPa)
CF3	70 (485)	30 (205)
CF3A	77 (530)	35 (240)
CF8	70 (485)	30 (205)
CF8A	77 (530)	35 (240)
CF3M	70 (485)	30 (205)
CF3MA	80 (550)	37 (255)
CF8M	70 (485)	30 (205)
CF3MN	75 (515)	37 (255)
CF8C	70 (485)	30 (205)
CF10	70 (485)	30 (205)
CF10M	70 (485)	30 (205)

CH8	65 (450)	28 (195)
CH10	70 (485)	30 (205)
CH20	70 (485)	30 (205)
CK20	65 (450)	28 (195)
HK30	65 (450)	35 (240)
HK40	62 (425)	35 (240)
HT30	65 (450)	28 (195)
CF10MC	70 (485)	30 (205)
CN7M	62 (425)	25 (170)
CN3MN	80 (550)	38 (260)
CD4MCu	100 (690)	70 (485)
CE8MN	95 (655)	65 (450)
CG8MMN	85 (585)	42.5 (295)
CG8M	75 (515)	35 (240)
CF10SMnN	85 (585)	42.5 (295)
CT15C	63 (435)	25 (170)
CK3MCuN	80 (550)	38 (260)
CE20N	80 (550)	40 (275)
CG3M	75 (515)	35 (240)
F45CD3M-WCuN	100 (700)	65 (450)

ASTM, A352/A352M-03 (Volume 01.02), Standard Specification for Steel Castings, Ferritic and Martensitic, for Pressure-Containing Parts, Suitable for Low-Temperature Service

Scope. This specification covers steel castings for valves, flanges, fittings, and other pressure-containing parts intended primarily for low-temperature service.

Several grades of ferritic steels and one grade of martensitic steel are covered. Selection of analysis depends on the design and service conditions.

Grade	Usual Minimum Testing Temperatures, °F (°C)
LCA	−25 (−32)
LCB	−50 (−46)
LCC	−50 (−46)
LC1	−75 (−59)
LC2	−100 (−73)

(Continues)

(Continued)

Grade	Usual Minimum Testing Temperatures, °F (°C)
LC2-1	−100 (−73)
LC3	−150 (−101)
LC4	−175 (−115)
LC9	−320 (−196)
CA6NM	−100 (−73)

Referenced Documents

ASTM

A351/A351M, Specification for Castings, Austenitic, Austenitic-Ferritic (Duplex), for Pressure-Containing Parts.
A370, Test Methods and Definitions for Mechanical Testing of Steel Products.
A488/A488M, Practice for Steel Castings, Welding, Qualifications of Procedures and Personnel.
A703/A703M, Specification for Steel Castings, General Requirements, for Pressure-Containing Parts.
E165, Test Method for Liquid Penetrant Examination.
E709, Guide for Magnetic Particle Examination.

MSS. SP-55, Quality Standard for Steel Castings for Valves, Flanges, and Fittings and Other Piping Components (Visual Method).

Sizes. Varies.

Methods of Manufacture.

Low alloy—open hearth, electric furnace, or basic oxygen.
Stainless steel—electric furnace, vacuum furnace, or one of the former followed by vacuum or electro slag-consumable remelting.

Heat Treatment. Refer to ASTM A352/A352M.

Chemical Requirements. Refer to ASTM A352/A352M.

Mechanical Requirements. These are extracted from ASTM A352/A352M:

Grade	Minimum Tensile Strength, ksi (MPa)	Minimum Yield Strength, ksi (MPa)
LCA	60.0–85.0 (415–585)	30.0 (205)
LCB	65.0–90.0 (450–620)	35.0 (240)
LCC	70.0–85.0 (485–655)	40.0 (275)
LC1	65.0–90.0 (450–620)	35.0 (240)
LC2	70.0–95.0 (485–655)	40.0 (275)
LC2–1	105.0–130.0 (725–895)	80.0 (550)
LC3	70.0–95.0 (485–655)	40.0 (275)
LC4	70.0–95.0 (485–655)	40.0 (275)
LC9	85.0 (585)	75.0 (515)
CA6NM	110.0–135.0 (760–930)	80.0 (550)

ASTM, A353/A353M-93(1999) (Volume 01.04), Standard Specification for Pressure Vessel Plates, Alloy Steel, 9% Nickel, Double-Normalized and Tempered

Scope. This specification covers 9% nickel steel plates, double-normalized and -tempered, intended particularly for welded pressure vessels in cryogenic service. Plates produced under this specification are subject to impact testing at −320°F (−195°C) or at such other temperatures as agreed on.

The maximum thickness of plates is limited only by the capacity of the material to meet the specific mechanical property requirements; however, current mill practice normally limits this material to 2 in. (50 mm) maximum.

This material is susceptible to magnetization. Use of magnets in handling after heat treatment should be avoided if residual magnetism would be detrimental to subsequent fabrication or service.

Referenced Documents

ASTM

A20/A20M, Specification for General Requirements for Steel Plates for Pressure Vessels.

A435/A435M, Specification for Straight-Beam Ultrasonic Examination of Steel Plates.

A577/A577M, Specification for Ultrasonic Angle-Beam Examination of Steel Plates.
A578/A578M, Specification for Straight-Beam Ultrasonic Examination of Plain and Clad Steel Plates for Special Applications.

Methods of Manufacture. The steel is killed and conforms to the fine austenitic grain size requirements of Specification of A20/A20M.

Heat Treatment. Refer to ASTM A353/A353M.

Chemical Requirements. Refer to ASTM A353/A353M.

Mechanical Requirements. These are extracted from ASTM A353/A353M:

Material	Grade	Tensile Strength, ksi (MPa)	Minimum Yield Strength, ksi (MPa)
A353	A	100–120 (690–825)	75 (515)

ASTM, A358/A358M-01 (Volume 01.01), Standard Specification for Electric-Fusion-Welded Austenitic Chromium-Nickel Alloy Steel Pipe for High-Temperature Service

Scope. This specification covers electric-fusion-welded austenitic chromium-nickel alloy steel pipe suitable for corrosive or high-temperature service, or both.

Note: The dimensionless designator NPS (nominal pipe size) has been substituted in this standard for such traditional terms as *nominal diameter*, *size*, and *nominal size*.

The selection of the proper alloy and requirements for heat treatment is at the discretion of the purchaser, depending on the service conditions to be encountered.

The five classes of pipe are covered as follows:

Class 1. Pipe double welded by processes employing filler metal in all passes and completely radiographed.
Class 2. Pipe double welded by processes employing filler metal in all passes. No radiography is required.

Class 3. Pipe single welded by processes employing filler metal in all passes and completely radiographed.

Class 4. Same as Class 3 except that the weld pass exposed to the inside pipe surface may be made without the addition of filler metal (see Specifications 6.2.2.1 and 6.2.2.2).

Class 5. Pipe double welded by processes employing filler metal in all passes and spot radiographed.

Referenced Documents

ASTM

A240/A240M, Specification for Heat-Resisting Chromium and Chromium-Nickel Stainless Steel Plate, Sheet, and Strip for Pressure Vessels.

A262, Practices for Detecting Susceptibility to Intergranular Attack in Austenitic Stainless Steels.

A480/A480M, Specification for General Requirements for Flat-Rolled Stainless and Heat-Resisting Steel Plate, Sheet, and Strip.

A941, Terminology Relating to Steel, Stainless Steel, Related Alloys, and Ferroalloys.

A999/A999M, Specification for General Requirements for Alloy and Stainless Steel Pipe.

E527, Practice for Numbering Metals and Alloys (UNS).

ASME BPV

Section I, Welding and Brazing Qualifications.

Section IX, Welding Qualifications.

AWS

A5.22, Flux Cored Arc Welding.

A5.30, Consumable Weld Inserts for Gas Tungsten Arc Welding.

A5.4, Corrosion-Resisting Chromium and Chromium-Nickel Steel Covered Welding Electrodes.

A5.9, Corrosion-Resisting Chromium and Chromium-Nickel Steel Welding Rods and Bare Electrodes.

A5.11, Nickel and Nickel-Alloy Covered Welding Electrodes.

A5.14, Nickel and Nickel-Alloy Bare Welding Rods and Electrodes.

SAE. J1086, Practice for Numbering Metals and Alloys (UNS).

Chemical Requirements. Refer to ASTM A358/A358M. The chemical composition of the plate conforms to the requirements of the applicable specification and grade listed in Specification A240.

Mechanical Requirements. These are extracted from ASTM A358/A358M. The plate used in making the pipe conforms to the requirements as to tensile properties of tensile properties listed in Specification A240.

ASTM, A403/A403M-03a (Volume 01.01), Standard Specification for Wrought Austenitic Stainless Steel Piping Fittings

Scope. This specification covers wrought stainless steel fittings for pressure piping applications. Several grades of austenitic stainless steel alloys are included in this specification. The grades are designated with a prefix, WP or CR, based on the applicable ASME or MSS dimensional and rating standards, respectively. For each of the WP stainless grades, several classes of fittings are covered, to indicate whether seamless or welded construction was utilized. Class designations are also utilized to indicate the non-destructive test method and extent of nondestructive examination (NDE).

Referenced Documents

ASTM

A262, Practices for Detecting Susceptibility to Intergranular Attack in Austenitic Stainless Steels.
A351/A351M, Specification for Castings, Austenitic, Austenitic-Ferritic (Duplex), for Pressure-Containing Parts.
A370, Test Methods and Definitions for Mechanical Testing of Steel Products.
A388/A388M, Practice for Ultrasonic Examination of Heavy Steel Forgings.
A480/A480M, Specification for General Requirements for Flat-Rolled Stainless and Heat-Resisting Steel Plate, Sheet, and Strip.

A743/A743M, Specification for Castings, Iron-Chromium, Iron-Chromium-Nickel, Corrosion-Resistant, for General Application.

A744/A744M, Specification for Castings, Iron-Chromium-Nickel, Corrosion-Resistant, for Severe Service.

A751, Test Methods, Practices, and Terminology for Chemical Analysis of Steel Products.

A960, Specification for Common Requirements for Wrought Steel Piping Fittings.

E112, Test Methods for Determining Average Grain Size.

E165, Test Method for Liquid Penetrant Examination.

E213, Practice for Ultrasonic Examination of Metal Pipe and Tubing.

ASME

ASME B16.9, Factory-made Wrought Steel Butt-Welding Fittings.

ASME B16.11, Forged Steel Fittings, Socket-Welding and Threaded.

ASME B16.28, Wrought Steel Butt-Welding Short Radius Elbows and Returns.

MSS

SP-25, Standard Marking System for Valves, Fittings, Flanges, and Unions.

SP-43, Standard Practice for Lightweight Stainless Steel Butt-Welding Fittings.

SP-79, Socket-Welding Reducer Inserts.

SP-95, Swaged(d) Nipples and Bull Plugs.

ASME BPV Code

Section VIII, Division I, Pressure Vessels.

Section IX, Welding Qualifications.

AWS

A5.4, Specification for Corrosion-Resisting Chromium and Chromium-Nickel Steel Covered Welding Electrodes.

A5.9, Specification for Corrosion-Resisting Chromium and Chromium-Nickel Steel Welding Rods and Bare Electrodes.

ANSI. SNT-TC-1A (1984), Recommended Practice for Nondestructive Testing Personnel Qualification and Certification.

Classes.

S (seamless)—no NDE.
W (welded)—radiography or ultrasonic testing.
WX (welded)—radiography.
WU (welded)—ultrasonic testing.

Sizes. Varies according to the applicable ASME or MSS dimensional range.

Methods of Manufacture. Shaping operations performed by hammering, pressing, piercing, upsetting, rolling, bending, fusion welding, machining, or a combination of two or more of these operations.

Heat Treatment. Refer to ASTM A403/A403M.

Welding Repair. Repair procedures and welder qualifications are in ASTM A488/A488M.

Chemical Requirements (maximum in percentages). Refer to ASTM A403/A403M.

Mechanical Requirements. These are extracted from ASTM A403/A403M:

Grade	Minimum Tensile Strength, ksi (MPa)	Minimum Yield Strength, ksi (MPa)
304, 304LN, 304H, 309, 310, 316, 316LN, 316H, 317, 317L, 321, 321H, 347, 347H, 348, 348H, S31725	75 (515)	30 (205)
304L, 316L	70 (485)	25 (170)
304N, 316N, S31726	80 (550)	35 (240)
XM-19	100 (690)	55 (380)
S31254	94–119 (650–820)	44 (300)
S33228	73 (500)	27 (185)
S34565	115 (795)	60 (415)

ASTM, A420/A420M-03 (Volume 01.01), Standard Specification for Piping Fittings of Wrought Carbon Steel and Alloy Steel for Low-Temperature Service

Scope. This specification covers wrought carbon steel and alloy steel fittings of seamless and welded construction, covered by the latest revision of ASME B16.9, ASME B16.11, ASME B16.28, MSS SP-79, and MSS SP-95. Fittings differing from these ASME and MSSVF standards should be furnished in accordance with Supplementary Requirement S6. These fittings are for use in pressure piping and pressure vessel service at low temperatures.

Referenced Documents

ASTM

A370, Test Methods and Definitions for Mechanical Testing of Steel Products.
A960, Specification for Common Requirements for Wrought Steel Piping Fittings.

ASME

B 16.9, Factory-made Wrought Steel Butt-Welding Fittings.
B 16.11, Forged Steel Fittings, Socket-Welding Threaded.
B 16.28, Wrought Steel Butt-Welding Short-Radius Elbows and Returns.

ASME BPV Code

Section VIII, Division 1, Pressure Vessels.
Section V, Nondestructive Examination.

MSS

SP-25, Standard Marking System for Valves, Fittings, Flanges, and Unions.
SP-79, Socket Welding Reducer Inserts.
SP-95, Swage(d) Nipples and Bull Plugs.

ANSI. ASNT (1984), Recommended Practice No. SNT-TC-1A.

Classes. W (welded)—radiography or ultrasonic testing.

Sizes. Varies according to the applicable ASME or MSS dimensional range.

Methods of Manufacture. Shaping operations performed by hammering, pressing, piercing, upsetting, working, bending, fusion welding, or a combination of two or more of these operations.

Heat Treatment. Refer to ASTM A420/A420M.

Welding Repair. Repair procedures and welder qualifications are in ASTM A488/A488M.

Chemical Requirements. Refer to ASTM A420/A420M.

Mechanical Requirements. These are extracted from ASTM A420/A420M:

Grade	Minimum Tensile Strength, ksi (MPa)	Minimum Yield Strength, ksi (MPa)
WPL6	60–85 (415–585)	35 (240)
WPL9	63–88 (435–610)	46 (315)
WPL3	65–90 (450–620)	35 (240)
WPL8	100–125 (690–865)	75 (515)

ASTM, A515/A515M-03 (Volume 01.04), Standard Specification for Pressure Vessel Plates, Carbon Steel, for Intermediate- and Higher-Temperature Service

Scope. This specification covers carbon-silicon steel plates primarily for intermediate- and higher-temperature service in welded boilers and other pressure vessels. Plates under this specification are available in three grades having different strength levels as follows:

Grade	Tensile Strength, ksi (MPa)
60	60–80 (415–550)
65	65–85 (450–585)
70	70–90 (485–620)

The maximum thickness of plates is limited only by the capacity of the composition to meet the specified mechanical property requirements; however, current practice normally limits the maximum thickness of plates furnished under this specification as follows:

Grade	Maximum Thickness, in. (mm)
60	8 (200)
65	8 (200)
70	8 (200)

Referenced Document.

ASTM. A20/A20M, Specification for General Requirements for Steel Plates for Pressure Vessels.

Sizes. Varies.

Methods of Manufacture. The steel is killed and made into a coarse austenitic grain size practice.

Heat Treatment. Plates 2 in. (50 mm) and under in thickness are normally supplied in the as-rolled condition. The plates may be ordered normalized, stress relieved, or both. Plates over 2 in. (50 mm) in thickness are normalized.

Chemical Requirements. Refer to ASTM A515/A515M.

Mechanical Requirements. These are extracted from ASTM A515/A515M:

Grade	Tensile Strength, ksi (MPa)	Minimum Yield Strength, ksi (MPa)
60	60–80 (415–550)	32 (220)
65	65–85 (450–585)	35 (240)
70	70–90 (485–620)	38 (260)

ASTM, A516/A516M-03 (Volume 01.04), Standard Specification for Pressure Vessel Plates, Carbon Steel, for Moderate- and Lower-Temperature Service

Scope. This specification covers carbon steel plates intended primarily for service in welded pressure vessels where improved notch toughness is important. Plates under this specification are available in four grades, having different strength levels as follows:

Grade	Tensile Strength, ksi (MPa)
55 (380)	55–75 (380–515)
60 (415)	60–80 (415–550)
65 (450)	65–85 (450–585)
70 (485)	70–90 (485–620)

The maximum thickness of plates is limited only by the capacity of the composition to meet the specified mechanical property requirements; however, current practice normally limits the maximum thickness of plates furnished under this specification as follows:

Grade	Maximum Thickness, in. (mm)
55 (380)	12 (305)
60 (415)	8 (205)
65 (450)	8 (205)
70 (485)	8 (205)

Referenced Documents

ASTM

A20/A20M, Specification for General Requirements for Steel Plates for Pressure Vessels.
A435/A435M, Specification for Straight-Beam Ultrasonic Examination of Steel Plates.
A577/A577M, Specification for Ultrasonic Angle-Beam Examination of Steel Plates.
A578/A578M, Specification for Straight-Beam Ultrasonic Examination of Plain and Clad Steel Plates for Special Applications.

Sizes. Varies.

Methods of Manufacture. The steel is killed and made to a coarse austenitic grain size, following the requirements of Specification ASTM A20/A20M.

Heat Treatment. Plates 1.5 in (40 mm) and under in thickness are normally supplied in the as rolled condition. The plates may be ordered normalized, or stress relieved, or both. Plates over 1.5 in. (40 mm) in thickness are normalized.

Chemical Requirements. Refer to ASTM A516/A516M.

Mechanical Requirements. These are extracted from ASTM A515/A515M:

Grade	Tensile Strength, ksi (MPa)	Minimum Yield Strength, ksi (MPa)
55	55–75 (380–515)	30 (205)
60	60–80 (415–550)	32 (220)
65	65–85 (450–585)	35 (240)
70	70–90 (485–620)	38 (260)

ASTM, A587-96 (2001) (Volume 01.01), Standard Specification for Electric-Resistance-Welded Low-Carbon Steel Pipe for the Chemical Industry

Scope. This specification covers electric-resistance-welded low-carbon steel pipe intended for use as process lines. Pipe ordered under this specification are suitable for severe forming operations involving flanging in all sizes and bending to close radii up to and including NPS 4.

Referenced Documents

ASTM

A53/A53M, Specification for Pipe, Steel, Black and Hot-Dipped, Zinc-Coated Welded and Seamless.

A370, Test Methods and Definitions for Mechanical Testing of Steel Products.

A530/A530M, Specification for General Requirements for Specialized Carbon and Alloy Steel Pipe.
A751, Test Methods, Practices, and Terminology for Chemical Analysis of Steel Products.
E213, Practice for Ultrasonic Inspection of Metal Pipe and Tubing.
E273, Practice for Ultrasonic Examination of Longitudinal Welded Pipe and Tubing.
E309, Practice for Eddy-Current Examination of Steel Tubular Products Using Magnetic Saturation.
E570, Practice for Flux Leakage Examination of Ferromagnetic Steel Tubular Products.

Methods of Manufacture. Refer to ASTM A 587.

Chemical Requirements. Refer to ASTM A 587.

Mechanical Requirements. These are extracted from ASTM A 587:

Grade	Minimum Tensile Strength, ksi (MPa)	Minimum Yield Strength, ksi (MPa)
A 587	48 (331)	30 (207)

ASTM, A671-96 (2001) (Volume 01.01), Standard Specification for Electric-Fusion-Welded Steel Pipe for Atmospheric and Lower Temperatures

Scope. This specification covers electric-fusion-welded steel pipe with filler metal added, fabricated from pressure-vessel-quality plate of several analyses and strength levels and suitable for high-pressure service at atmospheric and lower temperatures. Heat treatment may or may not be required to attain the desired properties or comply with applicable code requirements. Supplementary requirements are provided for use when additional testing or examination is desired.

The specification nominally covers pipe 16 in. (405 mm) in outside diameter or larger and of $\frac{1}{4}$ in. (6.4 mm) wall thickness or greater. Pipe having other dimensions may be furnished provided it complies with all other requirements of this specification.

Class	Heat Treatment on Pipe	Radiography (see Section)	Pressure Test (see Section)
10	None	None	None
11	None	9	None
12	None	9	8.3
13	None	None	8.3
20	Stress relieved, see 5.3.1	None	None
21	Stress relieved, see 5.3.1	9	None
22	Stress relieved, see 5.3.1	9	8.3
23	Stress relieved, see 5.3.1	None	8.3
30	Normalized, see 5.3.2	None	None
31	Normalized, see 5.3.2	9	None
32	Normalized, see 5.3.2	9	8.3
33	Normalized, see 5.3.2	None	8.3
40	Normalized and tempered, see 5.3.3	None	None
41	Normalized and tempered, see 5.3.3	9	None
42	Normalized and tempered, see 5.3.3	9	8.3
43	Normalized and tempered, see 5.3.3	None	8.3
50	Quenched and tempered, see 5.3.4	None	None
51	Quenched and tempered, see 5.3.4	9	None
52	Quenched and tempered, see 5.3.4	9	8.3
53	Quenched and tempered, see 5.3.4	None	8.3
60	Normalized and precipitation heat treated	None	None
61	Normalized and precipitation heat treated	9	None
62	Normalized and precipitation heat treated	9	8.3
63	Normalized and precipitation heat treated	None	8.3
70	Quenched and precipitation heat treated	None	None
71	Quenched and precipitation heat treated	9	None
72	Quenched and precipitation heat treated	9	8.3
73	Quenched and precipitation heat treated	None	8.3

Note: Selection of materials should be made with attention to temperature of service. For such guidance, Specification A20/A20M may be consulted.

Referenced Documents.

ASTM

A20/A20M, Specification for General Requirements for Steel Plates for Pressure Vessels.
A370, Test Methods and Definitions for Mechanical Testing of Steel Products.
A435/A435M, Specification for Straight-Beam Ultrasonic Examination of Steel Plates.
A530/A530M, Specification for General Requirements for Specialized Carbon and Alloy Steel Pipe.
A577/A577M, Specification for Ultrasonic Angle-Beam Examination of Steel Plates.
A578/A578M, Specification for Straight-Beam Ultrasonic Examination of Plain and Clad Steel Plates for Special Applications.
E110, Test Method for Indentation Hardness of Metallic Materials by Portable Hardness Testers.
E165, Test Method for Liquid Penetrant Inspection.
E350. Test Method for Chemical Analysis of Carbon Steel, Low-Alloy Steel, Silicon Electrical Steel, Ingot Iron, and Wrought Iron.
E709. Practice for Magnetic Particle Examination.
A203/A203M. Specification for Pressure Vessel Plates, Alloy Steel, Nickel.
A285/A285M. Specification for Pressure Vessel Plates, Carbon Steel, Low- and Intermediate-Tensile Strength.
A299/A299M. Specification for Pressure Vessel Plates, Carbon Steel, Manganese-Silicon.
A353/A353M. Specification for Pressure Vessel Plates, Alloy Steel, 9% Nickel, Double-Normalized and Tempered.
A515/A515M, Specification for Pressure Vessel Plates, Carbon Steel, for Intermediate-and Higher-Temperature Service.
A516/A516M, Specification for Pressure Vessel Plates, Carbon Steel, for Moderate- and Lower-Temperature Service.
A517/A517M, Specification for Pressure Vessel Plates, Alloy Steel, High-Strength, Quenched and Tempered.
A537/A537M, Specification for Pressure Vessel Plates, Heat-Treated, Carbon-Manganese-Silicon Steel.
A553/A553M, Specification for Pressure Vessel Plates, Alloy Steel, Quenched and Tempered 8% and 9% Nickel.
A645/A645M, Specification for Pressure Vessel Plates, 5% Nickel Alloy Steel, Specially Heat Treated.

A736/A736M, Specification for Pressure Vessel Plates, Low-Carbon Age-Hardening, Nickel-Copper-Chromium-Molybdenum-Columbium and Nickel-Copper-Manganese-Molybdenum-Columbium Alloy Steel.

A442/A442M, Specification for Pressure Vessel Plates, Carbon Steel, Improved Transition Properties.

ASME BPV Code

Section II, Material Specifications.
Section III, Nuclear Vessels.
Section VIII, Unfired Pressure Vessels.
Section IX, Welding Qualifications.

Heat Treatment. Refer to ASTM A671/A671M.

Plate Specifications. These are extracted from ASTM A671/A671M:

Grade	Type of Steel	Plate ASTM	Grade
CA 55	Plain carbon	A285/A285M	C
CB 60	Plain carbon, killed	A515/A515M	60
CB 65	Plain carbon, killed	A515/A515M	65
CB 70	Plain carbon, killed	A515/A515M	70
CC 60	Plain carbon, killed, fine grain	A516/A516M	60
CC 65	Plain carbon, killed, fine grain	A516/A516M	65
CC 70	Plain carbon killed, fine grain	A516/A516M	70
CD 70	Manganese-silicon, normalized	A537/A537M	1
CD 80	Manganese-silicon, normalized and tempered	A537/A537M	2
CE 55	Plain carbon	A442/A442M	55
CE 60	Plain carbon	A442/A442M	60
CF 65	Nickel steel	A203/A203M	A
CF 70	Nickel steel	A203/A203M	B
CF 66	Nickel steel	A203/A203M	D
CF 71	Nickel steel	A203/A203M	E
CJ 101	Alloy steel, quenched and tempered	A517/A517M	A
CJ 102	Alloy steel, quenched and tempered	A517/A517M	B
CJ 103	Alloy steel, quenched and tempered	A517/A517M	C
CJ 104	Alloy steel, quenched and tempered	A517/A517M	D
CJ 105	Alloy steel, quenched and tempered	A517/A517M	E
CJ 106	Alloy steel, quenched and tempered	A517/A517M	F
CJ 107	Alloy steel, quenched and tempered	A517/A517M	G

(Continues)

(Continued)

Grade	Type of Steel	Plate ASTM	Grade
CJ 108	Alloy steel, quenched and tempered	A517/A517M	H
CJ 109	Alloy steel, quenched and tempered	A517/A517M	J
CJ 110	Alloy steel, quenched and tempered	A517/A517M	K
CJ 111	Alloy steel, quenched and tempered	A517/A517M	L
CJ 112	Alloy steel, quenched and tempered	A517/A517M	M
CJ 113	Alloy steel, quenched and tempered	A517/A517M	P
CK 75	Carbon-manganese-silicon	A299/A299M	
CP 65	Alloy steel, age hardened, normalized, and precipitation heat treated	A517/A517M	2
CP 75	Alloy steel, age hardened, normalized, and precipitation heat treated	A517/A517M	3

Mechanical Requirements. These are extracted from ASTM A671/A671M. The plate used in making the pipe shall conform to the requirements as to tensile properties of various ASTM specifications listed in the preceding table.

ASTM, A672-96 (2001) (Volume 01.01), Standard Specification for Electric-Fusion-Welded Steel Pipe for High-Pressure Service at Moderate Temperatures

Scope. This specification covers steel pipe: electric-fusion-welded with filler metal added, fabricated from pressure-vessel-quality plate of any of several analyses and strength levels and suitable for high-pressure service at moderate temperatures. Heat treatment may or may not be required to attain the desired properties or comply with applicable code requirements. Supplementary requirements are provided for use when additional testing or examination is desired.

The specification nominally covers pipe 16 in. (405 mm) in outside diameter or larger with wall thicknesses up to 3 in. (75 mm), inclusive. Pipe having other dimensions may be furnished provided it complies with all other requirements of this specification.

Several grades and classes of pipe are provided. The grade designates the type of plate used. The class designates the type of heat treatment performed during manufacture of the pipe, whether the weld is radiographically examined, and whether the pipe has been pressure tested as listed.

Referenced Documents

ASTM

A20/A20M, Specification for General Requirements for Steel Plates for Pressure Vessels.
A370, Test Methods and Definitions for Mechanical Testing of Steel Products.
A435/A435M, Specification for Straight-Beam Ultrasonic Examination of Steel Plates.
A530/A530M, Specification for General Requirements for Specialized Carbon and Alloy Steel Pipe.
A577/A577M, Specification for Ultrasonic Angle-Beam Examination of Steel Plates.
A578/A578M, Specification for Straight-Beam Ultrasonic Examination of Plain and Clad Steel Plates for Special Applications.
E110, Test Method for Indentation Hardness of Metallic Materials by Portable Hardness Testers.
E165, Test Method for Liquid Penetrant Examination.
E350, Test Methods for Chemical Analysis of Carbon Steel, Low-Alloy Steel, Silicon Electrical Steel, Ingot Iron, and Wrought Iron.
E709, Guide for Magnetic Particle Examination.
A202/A202M, Pressure Vessel Plates, Alloy Steel, Chromium-Manganese-Silicon.
A204/A204M, Pressure Vessel Plates, Alloy Steel, Molybdenum.
A285/A285M, Pressure Vessel Plates, Carbon Steel, Low and Intermediate Tensile Strength.
A299/A299M, Pressure Vessel Plates, Carbon Steel, Manganese-Silicon.
A302/A302M, Pressure Vessel Plates, Alloy Steel, Manganese-Molybdenum and Manganese-Molybdenum-Nickel.
A515/A515M, Pressure Vessel Plates, Carbon Steel, for Intermediate- and Higher-Temperature Service.
A516/A516M, Pressure Vessel Plates, Carbon Steel, for Moderate- and Lower-Temperature Service.
A533/A533M, Pressure Vessel Plates, Alloy Steel, Quenched and Tempered, Manganese-Molybdenum and Manganese-Molybdenum-Nickel.
A537/A537M, Pressure Vessel Plates, Heat-Treated, Carbon-Manganese-Silicon Steel.
A442/A442M, Pressure Vessel Plates, Carbon Steel, Improved Transition Properties.

ASME BPV Code

Section II, Material Specifications.
Section III, Nuclear Vessels.
Section VIII, Unfired Pressure Vessels.
Section IX, Welding Qualifications.

Methods of Manufacture. The joints are double welded, with full penetration welds, made either manually or automatically and in accordance with standard procedures and by welders qualified in accordance with the ASME Boiler and Pressure Vessel Code, Section IX.

Heat Treatment. Refer to ASTM A672/A672M.

Plate Specifications. These are extracted from ASTM A672/A672M:

Grade	Type of Steel	Plate ASTM	Grade
A 45	Plain carbon	A285/A285M	A
A 50	Plain carbon	A285/A285M	B
A 55	Plain carbon	A285/A285M	C
B 55	Plain carbon, killed	A515/A515M	55
B 60	Plain carbon, killed	A515/A515M	60
B 65	Plain carbon, killed	A515/A515M	65
B 70	Plain carbon, killed	A515/A515M	70
C 55	Plain carbon, killed, fine grain	A516/A516M	55
C 60	Plain carbon, killed, fine grain	A516/A516M	60
C 65	Plain carbon, killed, fine grain	A516/A516M	65
C 70	Plain carbon, killed, fine grain	A516/A516M	70
D 70	Manganese-silicon, normalized	A537/A537M	1
D 80	Manganese-silicon, Q&T	A537/A537M	2
E 55	Plain carbon	A442/A442M	55
E 60	Plain carbon	A442/A442M	60
H 75	Manganese-molybdenum normalized	A302/A302M	A
H 80	Manganese-molybdenum normalized	A302/A302M	B, C, D
J 80	Manganese-molybdenum, quenched and tempered	A533/A533M	Cl-1
J 90	Manganese-molybdenum, quenched and tempered	A533/A533M	Cl-2
J 100	Manganese-molybdenum, quenched and tempered	A533/A533M	Cl-3
K 75	Chromium-manganese-silicon	A202/A202M	A
K 85	Chromium-manganese-silicon	A202/A202M	B
L 65	Molybdenum	A204/A204M	A
L 70	Molybdenum	A204/A204M	B
L 75	Molybdenum	A204/A204M	C
N 75	Manganese-silicon	A299/A299M	

Tensile Requirements (ASTM A672/A672M). The plate used in making the pipe conforms to the requirements as to tensile properties of various ASTM specifications listed in the preceding table.

ASTM, A691-98 (2002) (Volume 01.01), Standard Specification for Carbon and Alloy Steel Pipe, Electric-Fusion-Welded for High-Pressure Service at High Temperatures

Scope. This specification covers carbon and alloy steel pipe, electric-fusion-welded with filler metal added, fabricated from pressure-vessel-quality plate of several analyses and strength levels and suitable for high-pressure service at high temperatures. Heat treatment may or may not be required to attain the desired mechanical properties or comply with applicable code requirements. Supplementary requirements are provided for use when additional testing or examination is desired.

The specification nominally covers pipe 16 in. (405 mm) in outside diameter and larger with wall thicknesses up to 3 in. (75 mm) inclusive. Pipe having other dimensions may be furnished provided it complies with all other requirements of this specification.

Class	Heat Treatment on Pipe	Radiography (see Section)	Pressure Test (see Section)
10	None	None	None
11	None	9	None
12	None	9	8.3
13	None	None	8.3
20	Stress relieved, see 5.3.1	None	None
21	Stress relieved, see 5.3.1	9	None
22	Stress relieved, see 5.3.1	9	8.3
23	Stress relieved, see 5.3.1	None	8.3
30	Normalized, see 5.3.2	None	None
31	Normalized, see 5.3.2	9	None
32	Normalized, see 5.3.2	9	8.3
33	Normalized, see 5.3.2	None	8.3
40	Normalized and tempered, see 5.3.3	None	None
41	Normalized and tempered, see 5.3.3	9	None
42	Normalized and tempered, see 5.3.3	9	8.3
43	Normalized and tempered, see 5.3.3	None	8.3
50	Quenched and tempered, see 5.3.4	None	None
51	Quenched and tempered, see 5.3.4	9	None
52	Quenched and tempered, see 5.3.4	9	8.3
53	Quenched and tempered, see 5.3.4	None	8.3

Note: Selection of materials should be made with attention to temperature of service. For such guidance, Specification A20/A20M may be consulted.

Referenced Documents

ASTM

A20/A20M, Specification for General Requirements for Steel Plates for Pressure Vessels.
A204/A204M, Specification for Pressure Vessel Plates, Alloy Steel, Molybdenum.
A299/A299M, Specification for Pressure Vessel Plates, Carbon Steel, Manganese-Silicon.
A370, Test Methods and Definitions for Mechanical Testing of Steel Products.
A387/A387M, Specification for Pressure Vessel Plates, Alloy Steel, Chromium-Molybdenum.
A435/A435M, Specification for Straight-Beam Ultrasonic Examination of Steel Plates.
A530/A530M, Specification for General Requirements for Specialized Carbon and Alloy Steel Pipe.
A537/A537M, Specification for Pressure Vessel Plates, Heat-Treated, Carbon-Manganese-Silicon Steel.
E165 Test Method for Liquid Penetrant Examination.
E709 Practice for Magnetic Particle Examination.

ASME BPV Code

Section II, Material Specifications.
Section III, Nuclear Power Plant Components.
Section VIII, Unfired Pressure Vessels.
Section IX, Welding Qualifications.

Grades. The grade designates the specification of the plate used to make the pipe.

Class. The class designates the heat treatment performed in the manufacture of the pipe.

Methods of Manufacture. Double welded, full penetration welds, made in accordance with the ASME Boiler and Pressure Vessel Code IX. The welds are made either manually or automatically by an electric process involving the deposition of filler metal.

Heat Treatment and Inspection. These are extracted from ASTM A691/A691M:

Class	Heat Treatment[a]	Radiography	Pressure Test
10	None	No	No
11	None	Yes	No
12	None	Yes	Yes
13	None	No	Yes
20	Stress relieved	No	No
21	Stress relieved	Yes	No
22	Stress relieved	Yes	Yes
23	Stress relieved	No	Yes
30	Normalized	No	No
31	Normalized	Yes	No
32	Normalized	Yes	Yes
33	Normalized	No	Yes
40	Normalized and tempered	No	No
41	Normalized and tempered	Yes	No
42	Normalized and tempered	Yes	Yes
43	Normalized and tempered	No	Yes
50	Quenched and tempered	No	No
51	Quenched and tempered	Yes	No
52	Quenched and tempered	Yes	Yes
53	Quenched and tempered	No	Yes

[a]For heat treatment parameters, refer to ASTM A691/691M.

Plate Materials. These are extracted from ASTM A691/691M:

Grade	Type of Steel	ASTM Specification	Grade	HB Maximum
CM-65	Carbon-molybdenum	A204/A204M	A	201
CM-70	Carbon-molybdenum	A204/A204M	B	201
CM-75	Carbon-molybdenum	A204/A204M	C	201
CMSH-70	Carbon-managanese-silicon steel, normalised	A537/A537M	1	
CMS-75	Carbon-managanese-silicon steel	A299/A299M		
CMSH-80	Carbon-managanese-silicon steel, quenched and tempered	A537/A537M	2	201
½ CR	½% chromium, ½% molybdenum steel	A387/A387M	2	201
1CR	1% chromium, ½% molybdenum steel	A387/A387M	12	201
1¼ CR	1¼ % chromium, ½% molybdenum steel	A387/A387M	11	201

(Continues)

(Continued)

Grade	Type of Steel	ASTM Specification	Grade	HB Maximum
2¼ CR	2¼% chromium, 1% molybdenum steel	A387/A387M	22	201
3CR	3% chromium, 1% molybdenum steel	A387/A387M	21	201
5CR	5% chromium, ½% molybdenum steel	A387/A387M	5	225
9CR	9% chromium, 1% molybdenum steel	A387/A387M	9	241
91	9% chromium, 1% molybdenum steel, vanadium, columbium	A387/A387M	91	241

ASTM, A790/A790M-03 (Volume 01.01), Standard Specification for Seamless and Welded Ferritic/Austenitic Stainless Steel Pipe

Scope. This specification covers seamless and straight-seam welded ferritic/austenitic steel pipe intended for general corrosive service, with particular emphasis on resistance to stress corrosion cracking. These steels are susceptible to embrittlement if used for prolonged periods at elevated temperatures.

Referenced Documents

ASTM

A370, Test Methods and Definitions for Mechanical Testing of Steel Products.
A941, Terminology Relating to Steel, Stainless Steel, Related Alloys and Feroalloys.
A999/A999M, Specification for General Requirements for Alloy and Stainless Steel Pipe.
E213, Practice for Ultrasonic Examination of Metal Pipe and Tubing.
E309, Practice for Eddy-Current Examination of Steel Tubular Products Using Magnetic Saturation.
E381, Method of Macroetch Testing Steel Bars, Billets, Blooms, and Forgings.
E426, Practice for Electromagnetic (Eddy-Current) Examination of Seamless and Welded Tubular Products, Austenitic Stainless Steel and Similar Alloys.
E527, Practice for Numbering Metals and Alloys (UNS).

ASME

B1.20.1, Pipe Threads, General Purpose.
B36.10, Welded and Seamless Wrought Steel Pipe.
B36.19, Stainless Steel Pipe.

SAE. J 1086, Practice for Numbering Metals and Alloys (UNS).

ANSI. SNT-TC-1A, Personal Qualification and Certification in Nondestructive Testing.

Grades. The grade designates the specification of the plate used to make the pipe.

Methods of Manufacture. Refer to ASTM A790/A790M.

Chemical Analysis. Refer to ASTM A790/A790M.

Mechanical Requirements. These are extracted from ASTM A790/A790M:

Grade	Minimum Tensile Strength, ksi (MPa)	Minimum Yield Strength, ksi (MPa)
S31803	90 (620)	65 (450)
S32205	90 (620)	65 (450)
S31500	92 (630)	64 (440)
S32550	110 (760)	80 (550)
S31200	100 (690)	65 (450)
S31260	100 (690)	65 (450)
S32304	87 (600)	58 (400)
S39274	116 (800)	80 (550)
S32750	116 (800)	80 (550)
S32760	109–130 (750–895)	80 (550)
S32900	90 (620)	70 (485)
S32950	100 (690)	70 (485)
S39277	120 (825)	90 (620)
S32520	112 (770)	80 (550)
S32906	116 (800) below 0.4 in.	96 (650)
	109 (750) 0.4 and above	80 (550)

ASTM, B127-98 (Volume 02.04), Standard Specification for Nickel-Copper Alloy (UNS N04400) Plate, Sheet, and Strip

Scope. This specification covers rolled nickel-copper alloy (UNS N04400) plate, sheet, and strip.

Referenced Documents

ASTM

B164, Specification for Nickel-Copper Alloy Rod, Bar, and Wire.
E8, Test Methods for Tension Testing of Metallic Materials.
E10, Test Method for Brinell Hardness of Metallic Materials.
E18, Test Methods for Rockwell Hardness and Rockwell Superficial Hardness of Metallic Materials.
E29, Practice for Using Significant Digits in Test Data to Determine Conformance with Specifications.
E112, Test Methods for Determining the Average Grain Size.
E140, Hardness Conversion Tables for Metals.
E76, Test Methods for Chemical Analysis of Nickel-Copper Alloys.
F155, Test Method for Temper of Strip and Sheet Metals for Electronic Devices (Spring-back Method).

MIL

MIL-STD-129, Marking for Shipment and Storage.
MIL-STD-271, Nondestructive Testing Requirements for Metals.

Mechanical Requirements. These are extracted from ASTM B127-98:

Grade (Plate)	Minimum Tensile Strength, ksi (MPa)	Minimum Yield Strength, ksi (MPa)
UNS04400 Annealed	70 (485)	28 (128)
UNS04400 As rolled	75 (515)	40 (275)

ASTM, B160-99 (Volume 02.04), Standard Specification for Nickel Rod and Bar

Scope. This specification covers nickel (UNS N02200) and low-carbon nickel (UNS N02201) in the form of hot-worked and cold-worked rod and bar.

Referenced Documents

ASTM

B162, Specification for Nickel Plate, Sheet, and Strip.
B880, General Requirements for Chemical Check Analysis of Nickel, Nickel Alloys, and Cobalt Alloys.
E8, Test Methods for Tension Testing of Metallic Materials.
E18, Test Methods for Rockwell Hardness and Rockwell Superficial Hardness of Metallic Materials.
E29, Practice for Using Significant Digits in Test Data to Determine Conformance with Specifications.
E140, Hardness Conversion Tables for Metals.
E39, Test Methods for Chemical Analysis of Nickel.

Mechanical Requirements. These are extracted from ASTM B160-93

Grade	Minimum Tensile Strength, ksi (MPa)	Minimum Yield Strength, ksi (MPa)
Cold worked		
Rounds 1 in. (25.4 mm) and less	80 (550)	60 (415)
Rounds over 1 in. (25.4 mm)	75 (515)	50 (345)
Square, hex, rectangle, all sizes	65 (450)	40 (275)
Hot worked		
All sections and sizes	60 (415)	15 (105)
Annealed		
Rods and bars, all sizes	55 (380)	145 (105)

ASTM, B162-99 (Volume 02.04), Standard Specification for Nickel Plate, Sheet, and Strip

Scope. This specification covers rolled nickel (UNS N02200) and low-carbon nickel (UNS N02201) plate, sheet, and strip. The values stated in inch-pound units are to be regarded as the standard. The other values given are for information only.

Referenced Documents

ASTM

B160, Specification for Nickel Rod and Bar.
B880, General Requirements for Chemical Check Analysis of Nickel, Nickel Alloys, and Cobalt Alloys.
E8, Test Methods for Tension Testing of Metallic Materials.
E10, Test Method for Brinell Hardness of Metallic Materials.
E18, Test Methods for Rockwell Hardness and Rockwell Superficial Hardness of Metallic Materials.
E29, Practice for Using Significant Digits in Test Data to Determine Conformance with Specifications.
E112, Test Methods for Determining the Average Grain Size.
E140, Hardness Conversion Tables for Metals.
E39, Test Methods for Chemical Analysis of Nickel.
F155, Test Method for Temper of Strip and Sheet Metals for Electronic Devices (Spring-back Method).

Mechanical Requirements. These are extracted from ASTM B162-99:

Grade (Plate)	Minimum Tensile Strength, ksi (MPa)	Minimum Yield Strength, ksi (MPa)
UNSN02200		
Anealed	55 (380)	15 (100)
As rolled	55 (380)	20 (135)
UNSN02201		
Anealed	50 (345)	12 (80)
As rolled	50 (345)	12 (80)

ASTM, B164-03 (Volume 02.04), Standard Specification for Nickel-Copper Alloy Rod, Bar, and Wire

Scope. This specification covers nickel-copper alloys UNS N04400 and N04405 in the form of hot-worked and cold-worked rod and bar.

Referenced Documents

ASTM

B127, Specification for Nickel-Copper Alloy (UNS N04400) Plate, Sheet, and Strip.
B880, Specification for General Requirements for Chemical Check Analysis Limits for Nickel, Nickel Alloys, and Cobalt Alloys.
E8, Test Methods for Tension Testing of Metallic Materials.
E18, Test Methods for Rockwell Hardness and Rockwell Superficial Hardness of Metallic Materials.
E29, Practice for Using Significant Digits in Test Data to Determine Conformance with Specifications.
E76, Test Methods for Chemical Analysis of Nickel-Copper Alloys.
E140, Hardness Conversion Tables for Metals.
E1473, Test Methods for Chemical Analysis of Nickel, Cobalt, and High-Temperature Alloys.

MIL

STD-129, Marking for Shipment and Storage.
STD-271, Nondestructive Testing Requirements for Metals.

Methods of Manufacture. Refer to ASTM B 164.

Heat Treatment. Refer to ASTM B 164.

Chemical Requirements. Refer to ASTM B 164.

Mechanical Requirements (for rod and bar). These are extracted from ASTM A 164:

Grade	Minimum Tensile Strength, ksi (MPa)	Minimum Yield Strength, ksi (MPa)
UNS N04400		
Cold worked (as worked):		
Rounds under ½ in. (12.7)	110 (760)	85 (585)
Squares, hexagons and rectangles under ½ in. (12.7)	85 (585)	55 (380)

(Continues)

(Continued)

Grade	Minimum Tensile Strength, ksi (MPa)	Minimum Yield Strength, ksi (MPa)
Cold worked (stress relieved):		
Rounds over 1/2 in. (12.7 mm)	84 (580)	50 (345)
Rounds ½–3½ in. (12.7–88.9 mm) inclusive	87 (600)	60 (415)
Rounds over 3½–4 in. (88.9–101.6 mm)	84 (580)	55 (380)
Squares, hexagons, rectangles 2 in. (50.8 mm) and under	84 (580)	50 (345)
Squares, hexagons, rectangles over 2–3⅛ in. (50.8–74.4 mm) inclusive	80 (552)	50 (345)
Hot worked (as worked or stress relieved):		
Rounds, squares, rectangles up to 12 in. (305 mm) inclusive	80 (552)	40 (276)
Rounds, squares, rectangles over 12–14 in. (305–356 mm)	75 (517)	40 (276)
Hexagons over 2⅛–4 in. (54–102 mm) inclusive	75 (517)	30 (207)
Rings and disks		
Hot or cold worked (annealed):	70 (480)	25 (170)
Rod and bar, all sizes		
Rings and disks		
UNS N04405		
Cold worked (as worked or stress relieved):		
Rounds under ½ in. (12.7 mm)	85 (585)	50 (345)
Rounds ½–3 in. (12.7–76.2 mm) inclusive	85 (585)	50 (345)
Rounds over 3–4 in. (76.2–101.6 mm)	80 (552)	50 (345)
Squares, hexagons 2 in. (50.8 mm) and under	85 (585)	50 (345)
Squares, hexagons, rectangles 2 (50.8 mm) to 3⅛ (79.4) inclusive		
Hot worked (as hot worked or stress relieved):		
Rounds 3 in. (76.2 mm) or less.	75 (517)	35 (241)
Hexagons, squares 2⅛ in. (54 mm) or less.	75 (517)	35 (241)
Hexagon, squares over 2⅛–4 in. (54–101.6 mm) inclusive	70 (480)	30 (207)
Hot or cold worked (annealed):	70 (480)	25 (170)
Rod and bar all sizes		

ASTM, B168-01 (Volume 02.04), Standard Specification for Nickel-Chromium-Iron Alloys (UNS N06600, N06601, N06603, N06690, N06693, N06025, and N06045) and Nickel-Chromium-Cobalt-Molybdenum Alloy (UNS N06617) Plate, Sheet, and Strip

Scope. This specification covers rolled nickel-chromium-iron alloys (UNS N06600, N06601, N06603, N06690, N06693, N06025, and N06045) and nickel-chromium-cobalt-molybdenum alloy (UNS N06617) plate, sheet, and strip. The values stated in inch-pound units are to be regarded as the standard. The values given in parentheses are for information only.

Referenced Documents

ASTM

B166, Specification for Nickel-Chromium-Iron Alloys (UNS N06600, N06601, N06603, N06690, N06693, N06025, and N06045) and Nickel-Chromium-Cobalt-Molybdenum Alloy (UNS N06617) Rod, Bar, and Wire.
B880, Specification for General Requirements for Chemical-Check-Analysis Limits for Nickel, Nickel Alloys, and Cobalt Alloys.
E8, Test Methods for Tension Testing of Metallic Materials.
E10, Test Method for Brinell Hardness of Metallic Materials.
E18, Test Methods for Rockwell Hardness and Rockwell Superficial Hardness of Metallic Materials.
E29, Practice for Using Significant Digits in Test Data to Determine Conformance with Specifications.
E112, Test Methods for Determining the Average Grain Size.
E140, Hardness Conversion Tables for Metals.
E1473, Test Methods for Chemical Analysis of Nickel, Cobalt, and High-Temperature Alloys.
E38, Methods for Chemical Analysis of Nickel-Chromium and Nickel-Chromium-Iron Alloys.
F155, Test Method for Temper of Strip and Sheet Metals for Electronic Devices (Spring-back Method).

Federal Standard

Fed. Std. No. 102, Preservation, Packaging, and Packing Levels.
Fed. Std. No. 123, Marking for Shipment (Civil Agencies).
Fed. Std. No. 182, Continuous Identification Marking of Nickel and Nickel-Base Alloys.

MIL. MIL-STD-129, Marking for Shipment and Storage.

Mechanical Requirements. Refer to ASTM B168-01.

ASTM, B333-03 (Volume 02.04), Standard Specification for Nickel-Molybdenum Alloy Plate, Sheet, and Strip

Scope. This specification covers plate, sheet, and strip of nickel-molybdenum alloys (UNS N10001, N10665, N10675, N10629, and N10624)

The following products are covered under this specification: Sheet and strip include hot- or cold-rolled, solution-annealed, and descaled products, unless the solution annealing is performed in an atmosphere yielding a bright finish. Plate includes hot- or cold-rolled, solution-annealed, and descaled products.

Referenced Documents

ASTM

B906, Specification for General Requirements for Flat-Rolled Nickel and Nickel Alloys Plate, Sheet, and Strip.
E112, Test Methods for Determining the Average Grain Size.

Mechanical Requirements. These are extracted from ASTM B333-03:

Grade (Plate)	Minimum Tensile Strength, ksi (MPa)	Minimum Yield Strength, ksi (MPa)
UNS N10001	100 (690)	45 (310)
UNS N10665	110 (760)	51 (350)

Grade (Plate)	Minimum Tensile Strength, ksi (MPa)	Minimum Yield Strength, ksi (MPa)
UNS N10675	110 (760)	51 (350)
UNS N10629	110 (760)	51 (350)

ASTM, B335-03 (Volume 02.04), Standard Specification for Nickel-Molybdenum Alloy Rod

Scope. This specification covers rods of nickel-molybdenum alloys (UNS N10001, N10665, N10675, N10629, and N10624).

The following products are covered under this specification: rods $\frac{5}{16}$–$\frac{1}{3}$ in. (7.94–19.05 mm), inclusive, in diameter, hot or cold finished, solution annealed, and pickled or mechanically descaled; and rods $\frac{3}{4}$–$3\frac{1}{2}$ in. (19.05–88.9 mm), inclusive, in diameter, hot or cold finished, solution annealed, and ground or turned.

Referenced Documents

ASTM

B880, Specification for General Requirements for Chemical-Check-Analysis Limits for Nickel, Nickel Alloys and Cobalt Alloys.
E8, Test Methods for Tension Testing of Metallic Materials.
E29, Practice for Using Significant Digits in Test Data to Determine Conformance with Specifications.
E55, Practice for Sampling Wrought Nonferrous Metals and Alloys for Determination of Chemical Composition.
E1473, Test Methods for Chemical Analysis of Nickel, Cobalt, and High-Temperature Alloys.

Methods of Manufacture. Refer to ASTM B335.

Heat Treatment. Refer to ASTM B335.

Chemical Requirements. Refer to ASTM B335.

Mechanical Requirements. These are extracted from ASTM B335.

Grade	Minimum Tensile Strength, ksi (MPa)	Minimum Yield Strength, ksi (MPa)
UNS N10001 5/8–1½ in. (7.94–38.1 mm) inclusive	115 (795)	46 (315)
Over 1½ -3½ in. (38.1–88.9 mm)		
N10665	100 (690)	46 (315)
N10675	112 (760)	51 (350)
N10629	110 (760)	51 (350)
N10624	104 (720)	46 (320)

ASTM, B387-90(2001) (Volume 02.04), Standard Specification for Molybdenum and Molybdenum Alloy Bar, Rod, and Wire

Scope. This specification covers unalloyed molybdenum and molybdenum alloy bars, rods, and wires as follows:

Molybdenum 360—unalloyed vacuum arc-cast molybdenum.
Molybdenum 361—unalloyed powder metallurgy molybdenum.
Molybdenum alloy 363—vacuum arc-cast molybdenum–0.5% titanium–0.1% zirconium (TZM) alloy.
Molybdenum alloy 364—powder metallurgy molybdenum–0.5% titanium–0.1% zirconium (TZM) alloy.
Molybdenum 365—unalloyed vacuum arc-cast molybdenum, low carbon.
Molybdenum alloy 366—vacuum arc-cast molybdenum, 30% tungsten alloy.

Referenced Documents

ASTM

E8, Test Methods for Tension Testing of Metallic Materials.
F289, Specification for Molybdenum Wire and Rod for Electronic Applications.

Mechanical Requirements. Refer to ASTM B387-90(2001).

ASTM, B409-01 (Volume 02.04), Standard Specification for Nickel-Iron-Chromium Alloy Plate, Sheet, and Strip

Scope. This specification covers UNS N08120, UNS N08890, UNS N08800, UNS N08810, and UNS N08811 in the form of rolled plate, sheet, and strip. Alloy UNS N08800 is normally employed in service temperatures up to and including 1100°F (593°C). Alloys UNS N08120, UNS N08810, UNS N08811, and UNS N08890 are normally employed in service temperatures above 1100°F (593°C) where resistance to creep and rupture is required, and they are annealed to develop controlled grain size for optimum properties in this temperature range.

Referenced Documents

ASTM

B408, Specification for Nickel-Iron-Chromium Alloy Rod and Bar.
B880, Specification of General Requirements for Chemical-Check-Analysis Limits for Nickel, Nickel Alloys, and Cobalt Alloys.
E8, Test Methods for Tension Testing of Metallic Materials.
E10, Test Method for Brinell Hardness of Metallic Materials.
E18, Test Methods for Rockwell Hardness and Rockwell Superficial Hardness of Metallic Materials.
E29, Practice for Using Significant Digits in Test Data to Determine Conformance with Specifications.
E112, Test Methods for Determining the Average Grain Size.
E140, Hardness Conversion Tables for Metals.
E1473, Test Methods for Chemical Analysis of Nickel, Cobalt, and High-Temperature Alloys.
F155 Test Method for Temper of Strip and Sheet Metals for Electronic Devices (Spring-back Method).

Mechanical Requirements. These are extracted from ASTM B409-01:

Grade (Plate)	Minimum Tensile Strength, ksi (MPa)	Minimum Yield Strength, ksi (MPa)
UNS N08120		
Annealed	90 (621)	40 (276)
UNS N08800		
Annealed	75 (520)	30 (205)
As rolled	80 (550)	35 (240)
UNS N08810		
Annealed	65 (450)	25 (170)
UNS N08811		
Annealed	65 (450)	25 (170)

ASTM, B424-98a (Volume 02.04), Standard Specification for Ni-Fe-Cr-Mo-Cu Alloy (UNS N08825 and UNS N08221) Plate, Sheet, and Strip

Scope. This specification covers rolled nickel-iron-chromium-molybdenum-copper alloy (UNS N08825 and UNS N08221) plate, sheet, and strip. The values stated in inch-pound units are to be regarded as the standard. The values given in parentheses are for information only.

Referenced Documents

ASTM

B425, Specification for Ni-Fe-Cr-Mo-Cu Alloy (UNS N08825 and UNS N08221) Rod and Bar.
B880, Specification of General Requirements for Chemical-Check-Analysis Limits for Nickel, Nickel Alloys, and Cobalt Alloys.
E8, Test Methods for Tension Testing of Metallic Materials.
E29, Practice for Using Significant Digits in Test Data to Determine Conformance with Specifications.
E1473, Test Methods for Chemical Analysis of Nickel, Cobalt, and High-Temperature Alloys.

Mechanical Requirements. These are extracted from ASTM B424-98a:

Grade	Minimum Tensile Strength, ksi (MPa)	Minimum Yield Strength, ksi (MPa)
Hot Rolled Plate:		
UNS N08825	85 (586)	35 (586)
UNS N08221	79 (544)	34 (235)
Cold Rolled Plate:		
UNS N08825	85 (586)	35 (586)
UNS N08221	79 (544)	34 (235)

ASTM, B435-03 (Volume 02.04), Standard Specification for UNS N06002, UNS N06230, UNS N12160, and UNS R30556 Plate, Sheet, and Strip

Scope. This specification covers alloys UNS N06002, UNS N06230, UNS N12160, and UNS R30556 in the form of rolled plate, sheet, and strip for heat-resisting and general corrosive service.

The following products are covered under this specification: Sheet and strip include hot- or cold-rolled, annealed, and descaled products, unless solution annealing is performed in an atmosphere yielding a bright finish. Plate includes hot-rolled, solution-annealed, and descaled products.

Referenced Document

ASTM

B906, Specification of General Requirements for Flat-Rolled Nickel and Nickel Alloys Plate, Sheet, and Strip.

Mechanical Requirements. These are extracted from ASTM B435-03:

Grade	Minimum Tensile Strength, ksi (MPa)	Minimum Yield Strength, ksi (MPa)
UNS N06002	95 (655)	35 (240)
UNS N06230	110 (750)	45 (310)
UNS R30556	100 (690)	45 (310)
UNS N12160	90 (670)	35 (240)

ASTM, B443-00e1 (Volume 02.04), Standard Specification for Nickel-Chromium-Molybdenum-Columbium Alloy (UNS N06625) and Nickel-Chromium-Molybdenum-Silicon Alloy (UNS N06219) Plate, Sheet, and Strip

Scope. This specification covers rolled nickel-chromium-molybdenum-columbium alloy (UNS N06625) and nickel-chromium-molybdenum-silicon alloy (UNS N06219) plate, sheet, and strip.

Alloy UNS N06625 products are furnished in two grades of different heat-treated conditions:

Grade 1 (annealed)—Material is normally employed in service temperatures up to 1100°F (593°C).

Grade 2 (solution annealed)—Material is normally employed in service temperatures above 1100°F (593°C) when resistance to creep and rupture is required.

Note: Hot-working or reannealing may change properties significantly, depending on working history and temperatures.)

Alloy UNS N06219 is supplied only in a solution-annealed condition.

Referenced Documents

ASTM

B446, Specification for Nickel-Chromium-Molybdenum-Columbium Alloy (UNS N06625) Rod and Bar.

B880, Specification of General Requirements for Chemical-Check-Analysis Limits for Nickel, Nickel Alloys, and Cobalt Alloys.

E8, Test Methods for Tension Testing of Metallic Materials.

E29, Practice for Using Significant Digits in Test Data to Determine Conformance with Specifications.

E354, Test Methods for Chemical Analysis of High-Temperature, Electrical, Magnetic, and Other Similar Iron, Nickel, and Cobalt Alloys.

E1473, Test Methods for Chemical Analysis of Nickel, Cobalt, and High-Temperature Alloys.

Mechanical Requirements. These are extracted from ASTM B443-00e1:

Grade	Minimum Tensile Strength, ksi (MPa)	Minimum Yield Strength, ksi (MPa)
UNS N06625 (plate)		
Grade 1 (annealed)	110 (758)	55 (379)
Grade 2 (solution annealed)	100 (690)	40 (276)

ASTM, B462-02 (Volume 02.04), Specification for Forged or Rolled UNS N06030, UNS N06022, UNS N06200, UNS N08020, UNS N08024, UNS N08026, UNS N08367, UNS N10276, UNS N10665, UNS N10675, and UNS R20033 Alloy Pipe Flanges, Forged Fittings, and Valves and Parts for Corrosive High-Temperature Service.

Scope. This specification covers forged or rolled UNS N06030, UNS N06022, UNS N06200, UNS N08020, UNS N08024, UNS N08026, UNS N08367, UNS N10276, UNS N10665, UNS N10675, and UNS R20033 pipe flanges, forged fittings, and valves and parts intended for corrosive high-temperature service.

Referenced Documents.

ASTM

A262, Practices for Detecting Susceptibility to Intergranular Attack in Austenitic Stainless Steels.

B472, Specification for UNS N06030, UNS N06022, UNS N06200, UNS N08020, UNS N08026, UNS N08024, UNS N08926, UNS N08367, UNS N10276, UNS N10665, UNS N10675, and UNS R20033 Nickel Alloy Billets and Bars for Reforging.

B880, Specification of General Requirements for Chemical-Check-Analysis Limits for Nickel, Nickel Alloys, and Cobalt Alloys.

E8, Test Methods for Tension Testing of Metallic Materials.

E1473, Test Methods for Chemical Analysis of Nickel, Cobalt, and High-Temperature Alloys.

E1916, Guide for the Identification and/or Segregation of Mixed Lots of Metals.

ASME. B16.5, Steel Pipe Flanges and Flanged Fittings (for applicable alloy UNS N08020).

MSS. SP-25, Standard Marking System for Valves, Fittings, Flanges, and Unions.

Methods of Manufacture. Refer to ASTM B462.

Heat Treatment. Refer to ASTM B462.

Chemical Requirements. Refer to ASTM B462.

Mechanical Requirements. These are extracted from ASTM B462:

Grade	Minimum Tensile Strength, ksi (MPa)	Minimum Yield Strength, ksi (MPa)
UNS N08020, UNS N08024, UNS N08026	80 (551)	35 (241)
UNS N08367	95 (655)	45 (310)
UNS NR20033	109 (750)	55 (3800)
UNS N06030	85 (586)	35 (241)
UNS N06022	100 (690)	45 (310)
UNS N06200	100 (690)	41 (283)
UNS N10276	100 (690)	41 (283)
UNS N10665	110 (760)	51 (350)
UNS N10675	110 (760)	51 (350)

ASTM, B463-04 (Volume 02.04), Standard Specification for UNS N08020, UNS N08026, and UNS N08024 Alloy Plate, Sheet, and Strip

Scope. This specification covers UNS N08020, UNS N08026, and UNS N08024 alloy plate, sheet, and strip.

Referenced Documents

ASTM

A262, Practices for Detecting Susceptibility to Intergranular Attack in Austenitic Stainless Steels.

B906, Specification of General Requirements for Flat-Rolled Nickel and Nickel Alloys Plate, Sheet, and Strip.

Mechanical Requirements. These are extracted from ASTM B463-04:

Grade	Minimum Tensile Strength, ksi (MPa)	Minimum Yield Strength, ksi (MPa)
UNS N08026	80 (551)	35 (241)
UNS N08020	80 (551)	35 (241)
UNS N08024	80 (551)	35 (241)

ASTM, B511-01 (Volume 02.04), Standard Specification for Nickel-Iron-Chromium-Silicon Alloy Bars and Shapes

Scope. This specification covers wrought alloys UNS N08330 and UNS N08332 in the form of hot-finished and cold-finished bar and shapes intended for heat-resisting applications and general corrosive service.

Referenced Documents

ASTM

B536 Specification for Nickel-Iron-Chromium-Silicon Alloys (UNS N08330 and N08332) Plate, Sheet, and Strip.
B880 Specification of General Requirements for Chemical-Check-Analysis Limits for Nickel, Nickel Alloys, and Cobalt Alloys.
E8 Test Methods for Tension Testing of Metallic Materials.
E29 Practice for Using Significant Digits in Test Data to Determine Conformance with Specifications.
E112 Test Methods for Determining the Average Grain Size.
E1473 Test Methods for Chemical Analysis of Nickel, Cobalt, and High-Temperature Alloys.

Heat Treatment. Refer to ASTM B511.

Chemical Requirements. Refer to ASTM B511.

Mechanical Requirements. These are extracted from ASTM B511:

Grade	Minimum Tensile Strength, ksi (MPa)	Minimum Yield Strength, ksi (MPa)
UNS N10003	70 (483)	30 (207)
UNS N10242	67 (462)	27 (186)

ASTM, B536-02 (Volume 02.04), Standard Specification for Nickel-Iron-Chromium-Silicon Alloys (UNS N08330 and N08332) Plate, Sheet, and Strip

Scope. This specification covers nickel-iron-chromium silicon alloys (UNS N08330 and UNS N08332) plate, sheet, and strip intended for heat resisting applications and general corrosive service.

Referenced Documents

ASTM

B880, Specification of General Requirements for Chemical-Check-Analysis Limits for Nickel, Nickel Alloys, and Cobalt Alloys.
E8, Test Methods for Tension Testing of Metallic Materials.
E10, Test Method for Brinell Hardness of Metallic Materials.
E18, Test Methods for Rockwell Hardness and Rockwell Superficial Hardness of Metallic Materials.
E29, Practice for Using Significant Digits in Test Data to Determine Conformance with Specifications.
E112, Test Methods for Determining the Average Grain Size.
E140, Hardness Conversion Tables for Metals.
E1473, Test Methods for Chemical Analysis of Nickel, Cobalt, and High-Temperature Alloys.

Mechanical Requirements. These are extracted from ASTM B536-02:

Grade	Minimum Tensile Strength, ksi (MPa)	Minimum Yield Strength, ksi (MPa)
UNS N08330	70 (483)	30 (207)
UNS N08332	67 (462)	27 (186)

ASTM, B564-00a (Volume 02.04), Standard Specification for Nickel Alloy Forgings

Scope. This specification covers forgings of nickel alloy UNS N02200; Ni-Cu alloy UNS N04400; Ni-Cr-Fe alloys UNS N06600, UNS N06603, and UNS N06690; Ni-Cr-Mo-Nb alloy UNS N06625; Ni-Cr-Mo-Si alloy UNS N06219; low-carbon Ni-Mo-Cr alloys UNS N10276 and UNS N06022; Ni-Cr-Mo-W alloy UNS N06110; low-carbon Ni-Cr-Mo-W alloy UNS N06686; Ni-Fe-Cr-Mo-Cu alloy UNS N08825; Fe-Ni-Cr-Mo-N alloy UNS N08367; low-carbon Ni-Cr-Mo alloy UNS N06058; low-carbon Ni-Cr-Mo alloy UNS N06059; low carbon Ni-Cr-Mo-Cu alloy UNS N06200; Ni-Mo-Cr-Fe alloy UNS N10242; Ni-Mo alloys UNS N10665 and UNS N10675; low-carbon Ni-Fe-Cr-Mo-Cu alloy UNS N08031; Ni-Cr-W-Mo alloy UNS N06230; Ni-Cr-Co-Mo alloy UNS N06617; Ni-Co-Cr-Si alloy UNS N12160; Ni-Fe-Cr alloys, Ni-Mo alloy UNS N10629, Ni-Cr-Fe-Al alloy UNS N06025; Ni-Cr-Fe-Si alloy UNS N06045; low-carbon Ni-Mo-Cr-Ta alloy UNS N06210; Ni-Mo-Cr-Fe alloy UNS N10624; and low-carbon Cr-Ni-Fe-N alloy UNS R20033.

The nickel-iron-chromium alloys are UNS N08120, UNS N08800, UNS N08810, and UNS N08811. Alloy UNS N08800 is normally employed in service temperatures up to and including 1100°F (593°C). Alloys UNS N08810, N08120, and UNS N08811 are normally employed in service temperatures above 1100°F where resistance to creep and rupture is required; these are annealed to develop controlled grain size for optimum properties in this temperature range.

Referenced Documents

ASTM

B880, Specification of General Requirements for Chemical-Check-Analysis Limits for Nickel, Nickel Alloys, and Cobalt Alloys.

E8, Test Methods for Tension Testing of Metallic Materials.

E29, Practice for Using Significant Digits in Test Data to Determine Conformance with Specifications.

E76, Test Methods for Chemical Analysis of Nickel-Copper Alloys.

E112, Test Methods for Determining the Average Grain Size.

E350, Test Methods for Chemical Analysis of Carbon Steel, Low-Alloy Steel, Silicon Electrical Steel, Ingot Iron, and Wrought Iron.

E1473, Test Methods for Chemical Analysis of Nickel, Cobalt, and High-Temperature Alloys.

MIL

MIL-STD-129, Marking for Shipment and Storage.
MIL-STD-271, Nondestructive Testing Requirements for Metals.

Methods of Manufacture. Refer to ASTM B564.

Heat Treatment. Refer to ASTM B564.

Chemical Requirements. Refer to ASTM B564.

Mechanical Requirements. These are extracted from ASTM A564:

Grade	Minimum Tensile Strength, ksi (MPa)	Minimum Yield Strength, ksi (MPa)
UNS N02200	55 (380)	15 (105)
UNS N04400	70 (483)	25 (172)
UNS N06600	80 (552)	35 (241)
UNS N06690	85 (586)	35 (241)
UNS N06058	110 (760)	52 (360)
UNS N06059	100 (690)	45 (310)
UNS N06200	100 (690)	41 (283)
UNS N08120	90 (621)	40 (276)
UNS N08800	75 (517)	30 (207)
UNS N08810 and UNS N08811	65 (448)	25 (172)
UNS N06625		
Up to 4 in. (102 mm)	120 (827)	60 (414)
Over 4 in. (102 mm)	110 (758)	50 (345)
UNS N 06110		
Up to 4 in. (102 mm)	95 (655)	45 (310)
Over 4 in. (102 mm)	90 (621)	40 (276)
UNS N08825	85 (586)	35 (241)
UNS N10276	100 (690)	41 (283)
UNS N06022	100 (690)	45 (310)
UNS N08367	95 (655)	45 (310)
UNS N08031	94 (650)	40 (276)
UNS N06230	110 (758)	45 (310)
UNS N06617	95 (655)	35 (241)
UNS N10665	110 (758)	51 (350)
UNS N10675	110 (758)	51 (350)
UNS N10242	105 (725)	45 (310)
UNS N06686	100 (690)	45 (310)
UNS N12160	90 (620)	35 (241)

Grade	Minimum Tensile Strength, ksi (MPa)	Minimum Yield Strength, ksi (MPa)
UNS R20033	109 (750)	55 (380)
UNS N10629	110 (758)	51 (350)
UNS N06025		
Up to 4 in. (102 mm)	98 (680)	45 (310)
Over 4 in. (102 mm)	84 (580)	45 (310)
UNS N06603	94 (650)	43 (300)
UNS N06045	90 (620)	35 (240)
UNS N10624	104 (720)	46 (320)
UNS N06210	100 (690)	45 (310)
UNS N06219	96 (660)	39 (270)

ASTM, B572-03 (Volume 02.04), Standard Specification for UNS N06002, UNS N06230, UNS N12160, and UNS R30556 Rod

Scope. This specification covers alloys UNS N06002, UNS N06230, UNS N12160, and UNS R30556 in the rod form for heat resisting and general-corrosive service.

The following products are covered under this specification: Rods $^{5}/_{16}$–$^{3}/_{4}$ in. (7.94–19.05 mm), inclusive, in diameter, hot or cold finished, solution annealed, and pickled or mechanically descaled; and rods $^{3}/_{4}$–3$^{1}/_{2}$ in. (19.05–88.9 mm), inclusive, in diameter, hot or cold finished, solution annealed, and ground or turned.

Referenced Documents

ASTM

B880, Specification of General Requirements for Chemical-Check-Analysis Limits for Nickel, Nickel Alloys, and Cobalt Alloys.

E8, Test Methods for Tension Testing of Metallic Materials.

E29, Practice for Using Significant Digits in Test Data to Determine Conformance with Specifications.

E55, Practice for Sampling Wrought Nonferrous Metals and Alloys for Determination of Chemical Composition.

E1473, Test Methods for Chemical Analysis of Nickel, Cobalt, and High-Temperature Alloys.

Heat Treatment. Refer to ASTM B572.

Chemical Requirements. Refer to ASTM B572.

Mechanical Requirements. These are extracted from ASTM B572:

Grade	Minimum Tensile Strength, ksi (MPa)	Minimum Yield Strength, ksi (MPa)
N06002	95 (660)	35 (240)
N06230	110 (760)	45 (310)
N12160	90 (620)	35 (240)
R30556	100 (690)	45 (310)

ASTM, B573-00 (Volume 02.04), Standard Specification for Nickel-Molybdenum-Chromium-Iron Alloy (UNS N10003, N10242) Rod

Scope. This specification covers nickel-molybdenum-chromium-iron alloy (UNS N10003) rod for use in general corrosive service.

The following products are covered under this specification: Rods $^{5}/_{16}$ –$^{3}/_{4}$ in. (7.94–19.05 mm), inclusive, in diameter, hot or cold finished, annealed, and pickled or mechanically descaled; and rods $^{3}/_{4}$–$3^{1}/_{2}$ in. (19.05–88.9 mm), inclusive, in diameter, hot or cold finished, annealed, and ground or turned.

Referenced Documents

ASTM

B880, Specification of General Requirements for Chemical-Check-Analysis Limits for Nickel, Nickel Alloys, and Cobalt Alloys.
E8, Test Methods for Tension Testing of Metallic Materials.
E29, Practice for Using Significant Digits in Test Data to Determine Conformance with Specifications.
E1473, Test Methods for Chemical Analysis of Nickel, Cobalt, and High-Temperature Alloys.

Heat Treatment. Refer to ASTM B573.

Chemical Requirements. Refer to ASTM B573.

Mechanical Requirements. These are extracted from ASTM A573:

Grade	Minimum Tensile Strength, ksi (MPa)	Minimum Yield Strength, ksi (MPa)
UNS N10003	100 (690)	40 (280)
UNS N10242	105 (725)	45 (310)

ASTM, B574-99a (Volume 02.04), Specification for Low-Carbon Nickel-Molybdenum-Chromium, Low-Carbon Nickel-Chromium-Molybdenum, Low-Carbon Nickel-Molybdenum-Chromium-Tantalum, Low-Carbon Nickel-Chromium-Molybdenum-Copper, Low-Carbon Nickel-Chromium-Molybdenum-Tungsten Alloy Rod

Scope. This specification covers rods of low-carbon nickel-molybdenum-chromium alloys (UNS N10276, N06022, and N06455), low-carbon nickel-chromium-molybdenum alloy (UNS N06059), and low-carbon nickel-chromium-molybdenum-tungsten (UNS N06686).

The following products are covered under this specification: Rods $^5/_{16}$–$^3/_4$ in. (7.94–19.05 mm), inclusive, in diameter, hot or cold finished, solution annealed, and pickled or mechanically descaled; and rods $^3/_4$–3$^1/_2$ in. (19.05–88.9 mm), inclusive, in diameter, hot or cold finished, solution annealed, and ground or turned.

Referenced Documents

ASTM

B880, Specification for General Requirements for Chemical-Check-Analysis Limits for Nickel, Nickel Alloys, and Cobalt Alloys.

E8, Test Methods for Tension Testing of Metallic Materials.

E29, Practice for Using Significant Digits in Test Data to Determine Conformance with Specifications.

E55, Practice for Sampling Wrought Nonferrous Metals and Alloys for Determination of Chemical Composition.

E1473, Test Methods for Chemical Analysis of Nickel, Cobalt, and High-Temperature Alloys.

Methods of Manufacture. Refer to ASTM B574.

Heat Treatment. Refer to ASTM B574.

Chemical Requirements. Refer to ASTM B574.

Mechanical Requirements. These are extracted from ASTM A574:

Grade	Minimum Tensile Strength, ksi (MPa)	Minimum Yield Strength, ksi (MPa)
N10276	100 (690)	41 (283)
N06022	100 (690)	45 (310)
N06455	100 (690)	40 (276)
N06059	100 (690)	52 (360)
N06058	110 (760)	45 (310)
N06200	100 (690)	41 (283)
N06210	100 (690)	45 (310)
N06686	100 (690)	45 (310)

ASTM, B575-04 (Volume 02.04), Standard Specification for Low-Carbon Nickel-Chromium-Molybdenum, Low-Carbon Nickel-Chromium-Molybdenum-Copper, Low-Carbon Nickel-Chromium-Molybdenum-Tantalum, and Low-Carbon Nickel-Chromium-Molybdenum-Tungsten Alloy Plate, Sheet, and Strip

Scope. This specification covers plate, sheet, and strip of low-carbon nickel-chromium-molybdenum alloys (UNS N10276, UNS N06022, UNS N06455, N06035, UNS N06058, and UNS N06059), low-carbon nickel-chromium-molybdenum-copper alloy (UNS N06200), low-carbon nickel-chromium-molybdenum-tantalum alloy (UNS N06210), and low-carbon nickel-chromium-molybdenum-tungsten alloy (UNS N06686).

The following products are covered under this specification: Sheet and strip include products hot or cold rolled, solution annealed, and descaled, unless solution annealing is performed in an atmosphere yielding a bright finish. Plate includes products hot or cold rolled, solution annealed, and descaled.

Referenced Documents

ASTM

B906, Specification of General Requirements for Flat-Rolled Nickel and Nickel Alloys Plate, Sheet, and Strip.
E112, Test Methods for Determining the Average Grain Size.
E140, Hardness Conversion Tables for Metals.

Mechanical Requirements. These are extracted from ASTM B575-04:

Grade	Minimum Tensile Strength, ksi (MPa)	Minimum Yield Strength, ksi (MPa)
UNS N10276	100 (690)	41 (283)
UNS N06022	100 (690)	45 (310)
UNS N06455	100 (690)	40 (276)
UNS N06059	100 (690)	45 (310)
UNS N06200	100 (690)	41 (283)
UNS N06686	100 (690)	45 (310)

ASTM, B672-02 (Volume 02.04), Standard Specification for Nickel-Iron-Chromium-Molybdenum-Columbium Stabilized Alloy (UNS N08700) Bar and Wire

Scope. This specification covers nickel-iron-chromium-molybdenum-columbium stabilized alloy (UNS N08700) bars.

Referenced Documents

ASTM

A262, Practices for Detecting Susceptibility to Intergranular Attack in Austenitic Stainless Steels.
B880, Specification for General Requirements for Chemical Check Analysis Limits for Nickel, Nickel Alloys and Cobalt Alloys.
E8, Test Methods for Tension Testing of Metallic Materials.
E29, Practice for Using Significant Digits in Test Data to Determine Conformance with Specifications.
E1473, Test Methods for Chemical Analysis of Nickel, Cobalt, and High-Temperature Alloys.

Heat Treatment. Refer to ASTM B672.

Chemical Requirements. Refer to ASTM B672.

Mechanical Requirements. These are extracted from ASTM A672:

Grade	Minimum Tensile Strength, ksi (MPa)	Minimum Yield Strength, ksi (MPa)
N08700	80 (550)	35 (240)

4

PIPING COMPONENTS

1. INTRODUCTION

This introduction covers the numerous dimensional standards that relate to the most commonly used piping components within a metallic piping system. A piping component is a fitting that does one or more of the following:

- Transports the fluid—pipe.
- Changes the direction of the flow—elbows, tee.
- Changes the size of the pipe—reducers, reducing tees, reducing couplings.
- Joins together pipe—flanges, couplings.
- Dismantles pipe—flanges, unions.
- Isolates the flow. Spectacle blinds, Spades and Spacers
- Reinforces branch connections—weldolets, threadolets, sockolets.

The term *piping component* is interchangeable with *piping fitting*. A piping fitting is considered a component of a piping system.

A piping system comprises a variety of these components, and they serve one or more function. For example, a reducing tee changes the direction of the flow and the size of the pipe. Each particular component is manufactured to a specific dimensional standard, with fixed tolerances, based on its size, pressure rating, the method of manufacture, and the choice of end connections. The table in the next section lists the piping components most commonly used with their purposes and the appropriate dimensional standards.

2. DIMENSIONAL STANDARDS OF PIPE

The most common reference dimensional standards for pipe are as follows:

Material Type	Construction	Size Range	Standard
Carbon steel	Seamless and welded	1/8–80 in.	AMSE B36.10
Stainless steel and other corrosion-resistant alloys	Seamless and welded	1/8–30 in.	ASME B36.19

3. DIMENSIONAL STANDARDS FOR PIPING COMPONENTS

The most commonly used piping components and the dimensional standards are as follows:

Type of Component	Function	Butt-Weld Ends	Threaded-Socket-Weld Ends	Held between Flanges
90° long radius (LR) elbow	Change direction	ASME B16.9 (1/2–48 in.)	ASME B16.11 (1/2–4 in.)	Not applicable
90° short radius (SR) elbow	Change direction	ASME B16.28 (1/2–48 in.)	Not applicable	Not applicable
45° Elbow	Change direction	ASME B16.9 (1/2–48 in.)	ASME B16.11 (1/2–4 in.)	Not applicable
180° return	Change direction	ASME B16.9 (1/2–48 in.)	Not applicable	Not applicable
Equal tee	Change direction	ASME B16.9 (1/2–48 in.)	ASME B16.11 (1/2–4 in.)	Not applicable

Type of Component	Function	Butt-Weld Ends	Threaded-Socket-Weld Ends	Held between Flanges
Reducing tee	Change direction and size	ASME B16.9 (½–48 in.)	ASME B16.11 (½–4 in.)	Not applicable
Reinforced branch (O'let)	Change direction and size	Manufacturer's standard	Manufacturer's standard	Not applicable
Eccentric reducer	Change size	ASME B16.9 (½–48 in.)	ASME B16.11 (½–4 in.)	Not applicable
Concentric reducer	Change size	ASME B16.9 (½–48 in.)	ASME B16.11 (½–4 in.)	Not applicable
Flanges	Join pipe and components	ASME B16.5 (½–48 in.)	ASME B16.5 (½–24 in.)	Not applicable
Flanges	Join pipe and components	ASME B16.47 (26–60 in.)	Not applicable	Not applicable
Couplings	Join pipe and components	Not applicable	ASME B16.11 (½–4 in.)	Not applicable
Unions	Join pipe and components	Not applicable	BS 3799	Not applicable
Spectacle blinds, spades and spacers	Isolation	Not applicable	Not applicable	API 590 or company's standards

Each piping component type also has one or more methods of being connected to pipe or another component. The end connection chosen can be selected from one of the flowing commonly used alternatives:

- Butt weld.
- Plain end or socket weld.
- Threading.
- Flanging.

Other, less commonly used methods include hubbed connections and SAE flanges, however the preceding four types cover a vast majority of end connections and on certain projects, all requirements.

Dimensional Standards Covering End Connections of Components

The most commonly used dimensional standards for end connections are as follows:

End Connection	Joint Type	ASME Standard	Size
Weld end (WE)	Butt weld	ASME B16.25	All sizes
Plain end (PE)	Socket weld	ASME B16.11	4 in. and below
Threaded (Thd)	Screwed	ASME B1.20.1	4 in. and below
Flanged (Flg)	Flanged	ASME B16.5	½–24 in.
Flanged (Flg)	Flanged	ASME B16.47	26–60 in.

Generally, a piping component has the same connection at both ends. However, it is possible to have a mixture, especially with valves; for example, flanged by threaded, flanged by socket weld, or threaded by socket weld. This is acceptable as long as both end connections satisfy the design conditions of the fluid being transported in the piping system.

As mentioned previously, numerous other national standards cover the dimensional standards for piping components, however, differences in the dimensions and tolerances, in a vast majority of cases, could make the components incompatible.

4. THE MANUFACTURE OF ELECTRIC RESISTANCE WELDED PIPE

We next examine the basic steps necessary to produce electric resistance welded (ERW) pipe.

Coil Feed Ramp

The coils are removed from storage and placed on the feed ramp (see Figure 4.1). Each coil is fed into the uncoiling unit.

First Forming Section

The roll transition section receives the product from the first forming section and continues the "rounding-up" process (see Figure 4.2).

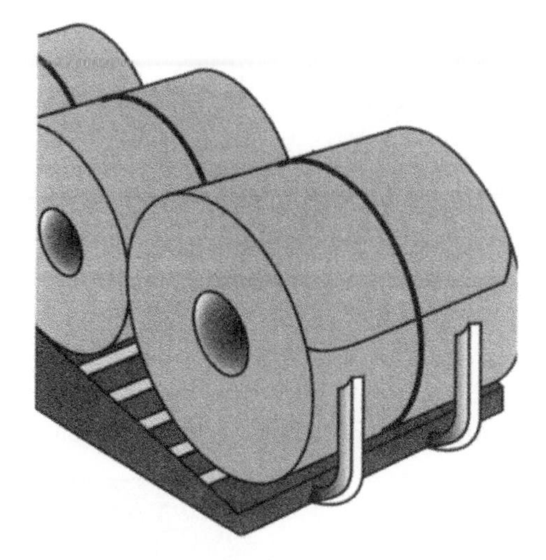

Figure 4.1. Coil Feed Ramp.

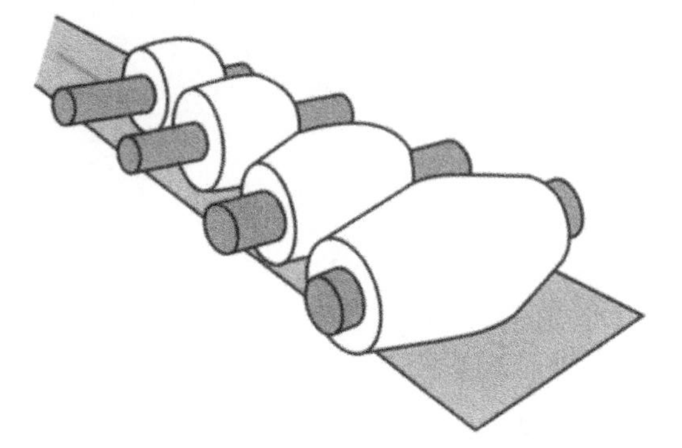

Figure 4.2. First Forming Section.

Final Forming Pass Section

This section of the forming rolls finishes the rounding process and prepares the edges of the strip for welding (see Figure 4.3).

High-Frequency Welder

An automatic high-frequency welder heats the edges of the strip to approximately 2600°F at the fusion point location (see Figure 4.4). Pressure rollers squeeze these heated edges together to form a fusion weld.

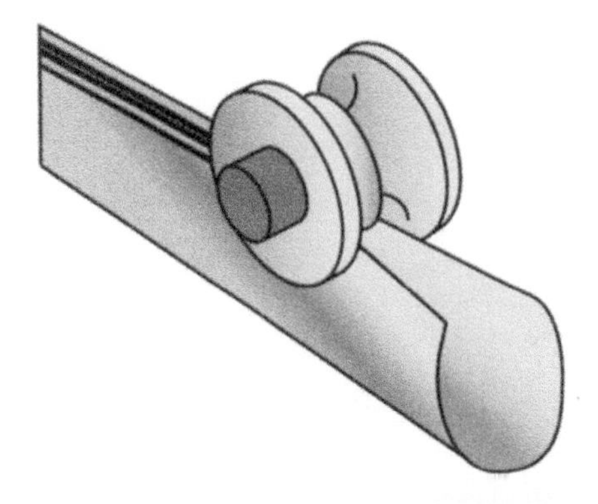

Figure 4.3. Final Forming Pass Section.

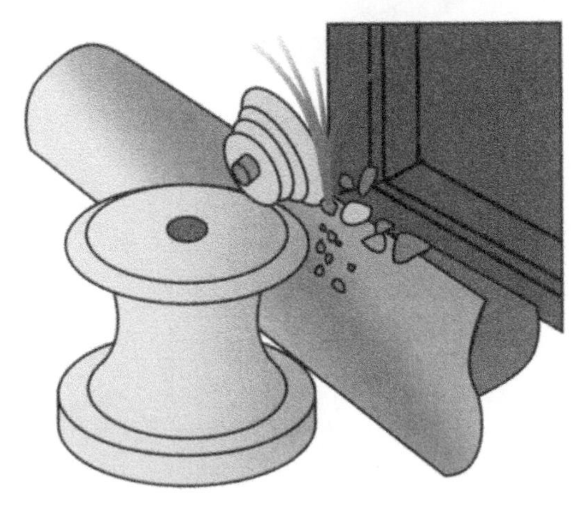

Figure 4.4. Automatic High-Frequency Welder.

In-Process Weld Nondestructive Inspection

After welding has been completed, it is inspected by independent nondestructive inspection units (see Figure 4.5).

Seam Weld Normalizing

The weld area is then subjected to postweld treatment, as metallurgically required, to remove residual welding stresses and produce a uniform normalized grain structure (see Figure 4.6).

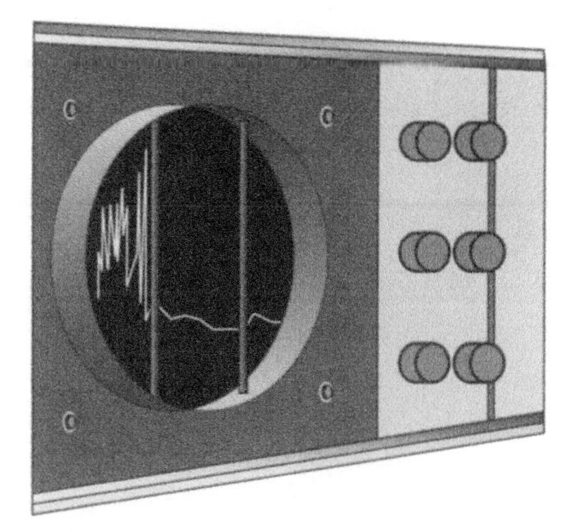

Figure 4.5. Nondestructive Inspection Unit.

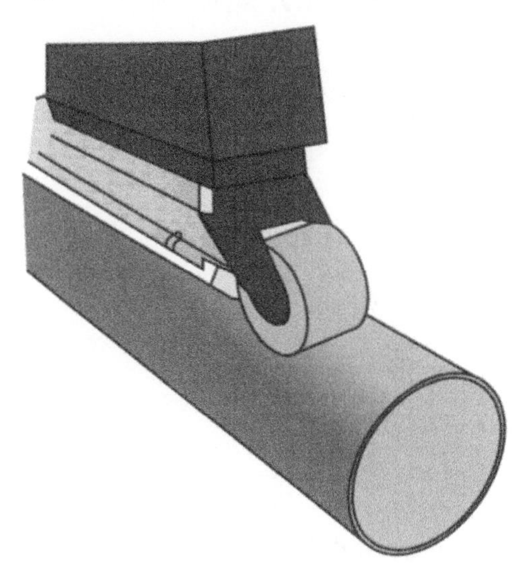

Figure 4.6. Seam Weld Normalizer.

Sizing Mill

The pipe passes through a sizing mill to achieve the correct outside diameter (see Figure 4.7).

Cutting the Pipe

The pipe is then cut to the correct length (see Figure 4.8).

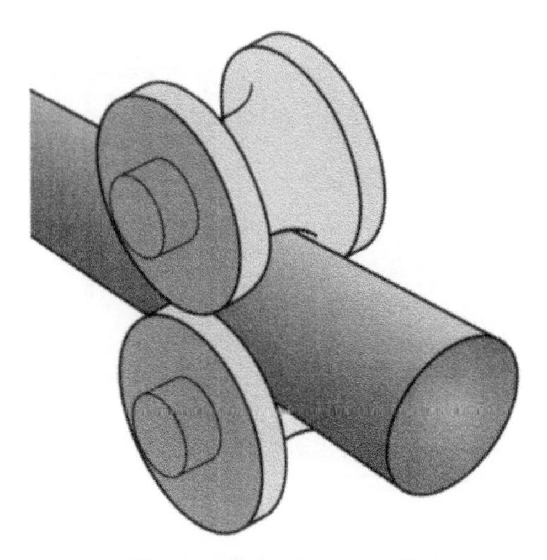

Figure 4.7. Sizing Mill.

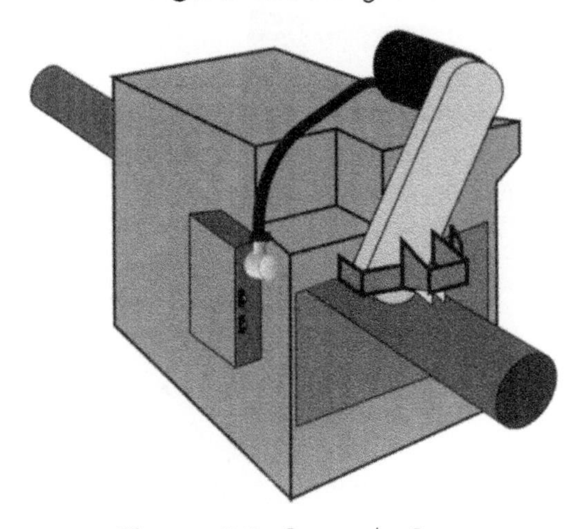

Figure 4.8. Cutting the Pipe.

Hydrostatic Testing

Each length of pipe is subjected to a hydrostatic test as a strength and leak check (see Figure 4.9).

Straightening

Each pipe length is then straightened by a series of horizontal deflection rolls (see Figure 4.10).

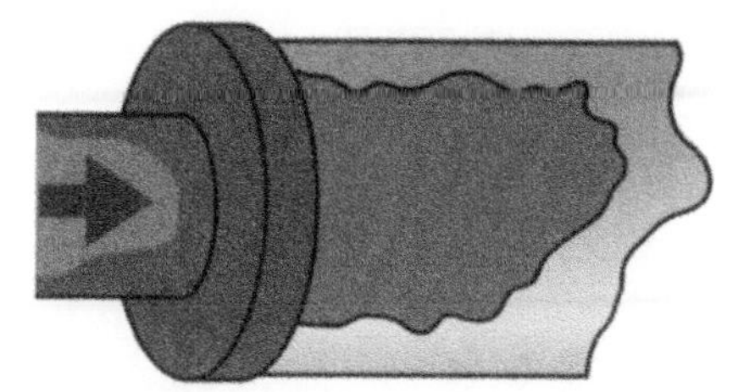

Figure 4.9. Hydrostatic Tester.

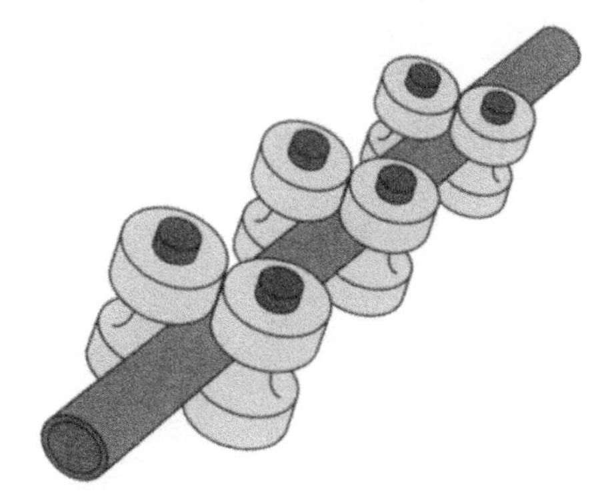

Figure 4.10. Horizontal Deflection Rolls.

Nondestructive Examination

The pipe is ultrasonically inspected and examined by electromagnetic means (see Figure 4.11).

Cutoff Facilities

Surface defects detected by nondestructive examination (NDE) are removed (see Figure 4.12).

Facing and Beveling

The desired end finish is then added to the pipe (see Figure 4.13).

Final Visual Inspection

A final visual inspection is given to the pipe prior to stenciling, loading, and shipping (see Figure 4.14).

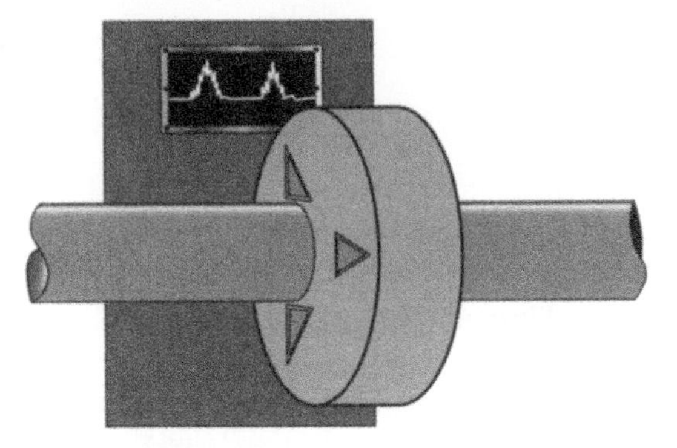

Figure 4.11. Electromagnetic Examination.

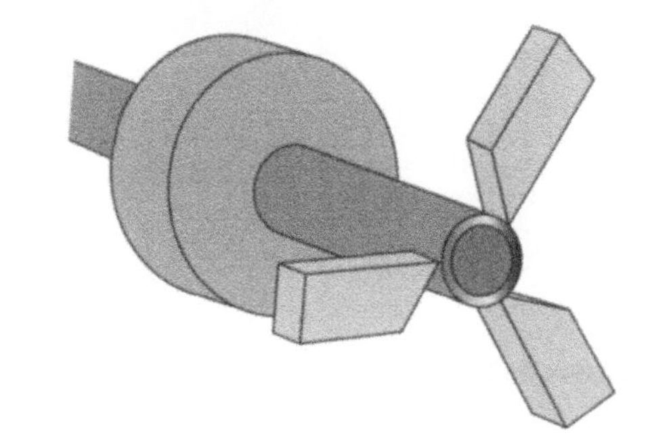

Figure 4.12. Removal of Surface Defects.

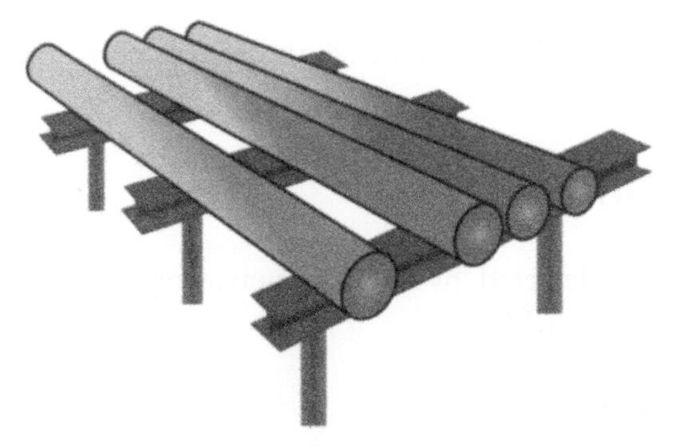

Figure 4.13. Adding the End Finish.

5. THE BASIC MANUFACTURE OF SEAMLESS PIPE

The methods of manufacturing seamless steel pipe vary slightly from manufacturer to manufacturer, but these are the basic stages.

Cast Round Billets

High-quality rounds are required for seamless tubular products (see Figure 4.14).

Round Reheating

The rounds are cut to the required length and weighed prior to being reheated in a furnace (see Figure 4.15).

Rotary Piercing Mill

The round billet is gripped by the rolls, which rotate and advance it into the piercer point, which creates a hole through its length (see Figure 4.16).

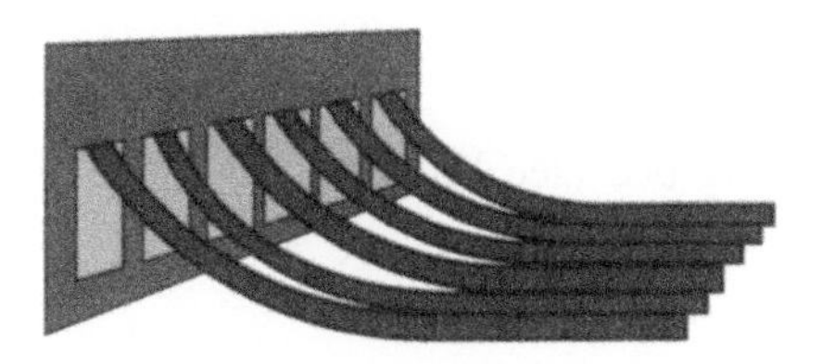

Figure 4.14. Casting Process.

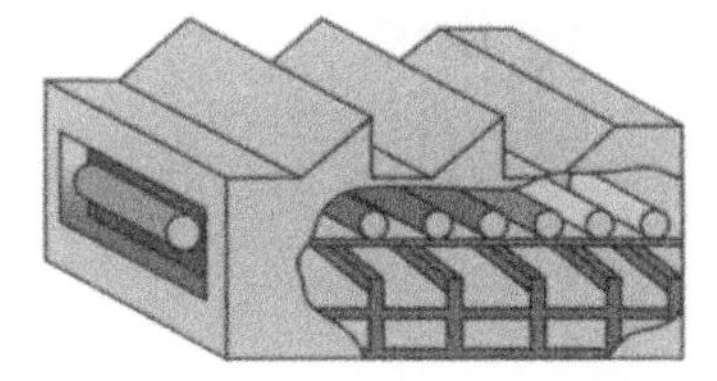

Figure 4.15. Round Reheating.

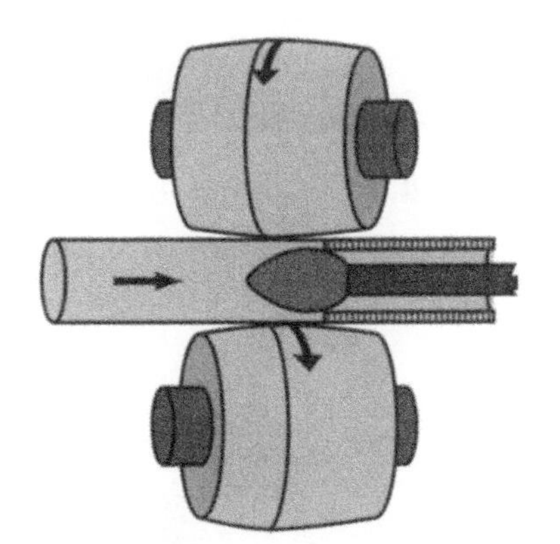

Figure 4.16. Rotary Piercing Mill (RPM).

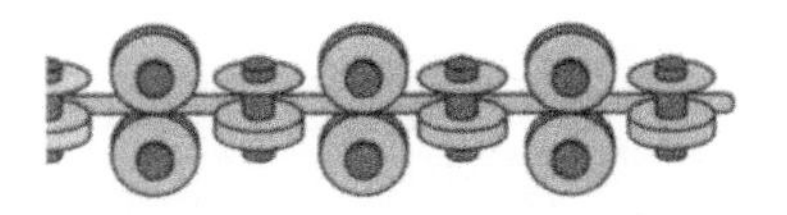

Figure 4.17. Mandrel Pipe Mill (MPM).

Mandrel Pipe Mill

The pipe is rolled using several stands over a long, restrained mandrel (see Figure 4.17).

Shell Reheating

The MPM shell is transferred to a reheat facility, where it can be cropped and weighed prior to reheating (see Figure 4.18).

Stretch Reducing Mill

The reheated and descaled pipe is conveyed through a stretch reducing mill, which utilizes up to 24 stands to reduce the diameter to the required finished size (see Figure 4.19).

Cooling Bed

The pipe lengths are placed on cooling bed (see Figure 4.20).

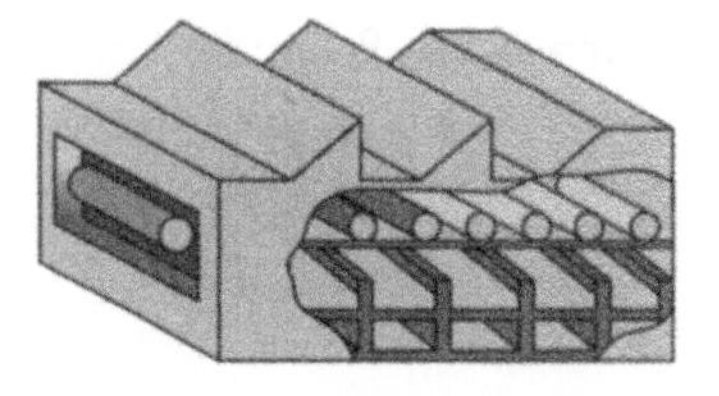

Figure 4.18. Shell Reheated in a Furnace.

Figure 4.20. Pipe on a Cooling Bed.

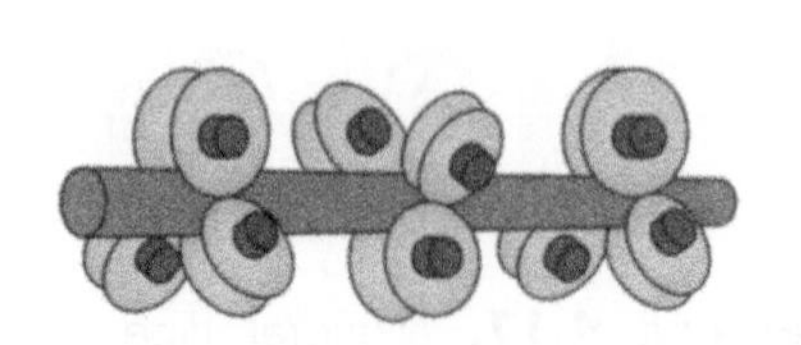

Figure 4.19. Stretch Reducing Mill.

Figure 4.21. Batch Saws.

Batch Saws

After cooling, batches of the as-rolled mother pipe are roller conveyed in parallel to carbide tipped batch saws for cropping into specified lengths (see Figure 4.21).

NDT Inspection

After cutting, the pipes are inspected for longitudinal and transverse flaws (electromagnetic), wall thickness (ultrasonic), and grade verification (eddy current) (see Figure 4.22).

Heat Treatment Furnace

Pipe to be heat treated can be austenitized in a walking beam furnace at a maximum temperature of about 1900°F (see Figure 4.23).

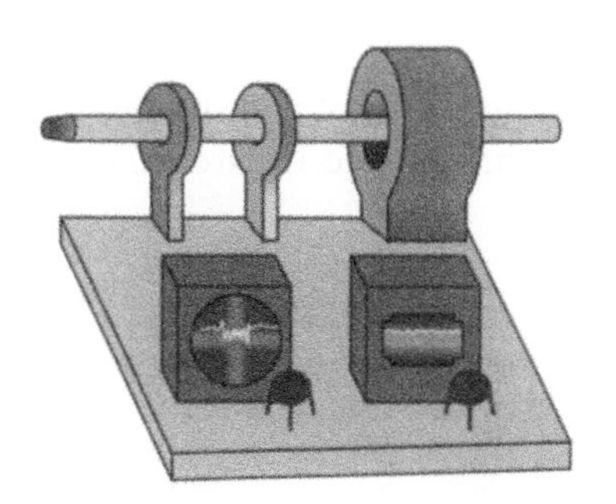

Figure 4.22. Nondestructive Testing.

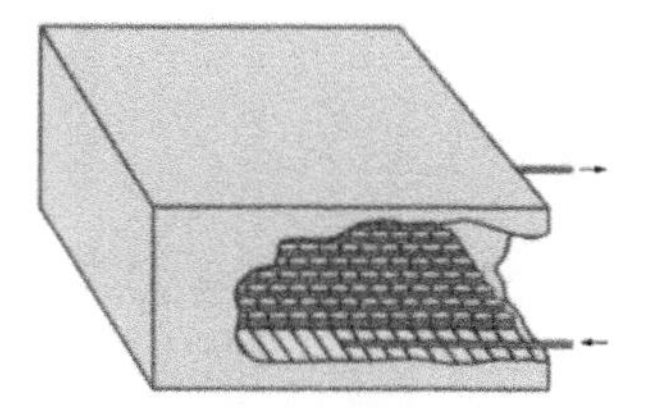

Figure 4.23. Walking Beam Furnace.

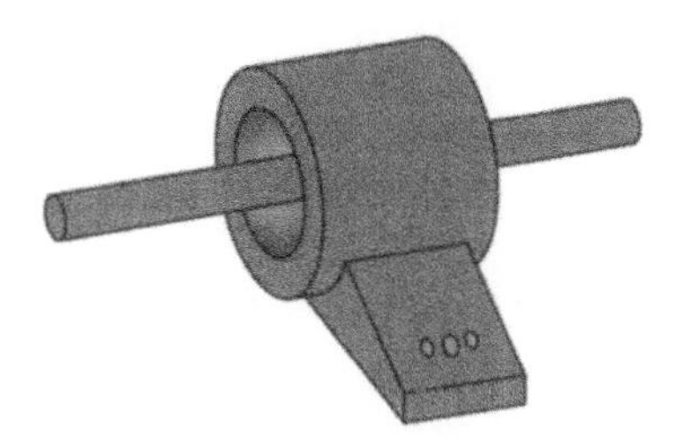

Figure 4.24. Quenching Unit.

Quenching Unit

On leaving the furnace, the hot pipe passes through a descaler and a three-section outside diameter water spray quencher (see Figure 4.24).

Tempering Furnace

The as-quenched pipe is tempered in a walking beam furnace at temperatures varying from 900°F to about 1300°F depending on grade (see Figure 4.25).

Sizing Mill

Pipe diameter tolerance is maintained by a three-stand, two-roll sizing mill (see Figure 4.26).

Hot Straightener

To ensure minimal effect on physical properties, all heat-treated pipe is straightened using a heated rotary straightener (see Figure 4.27).

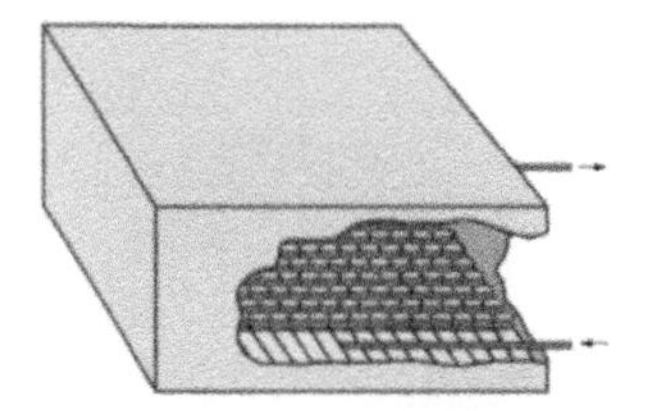

Figure 4.25. Tempering Furnace.

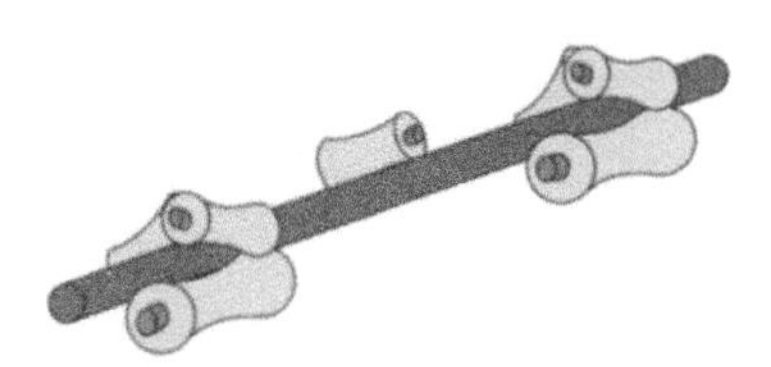

Figure 4.27. Rotary Straightener.

Figure 4.26. Sizing Mill.

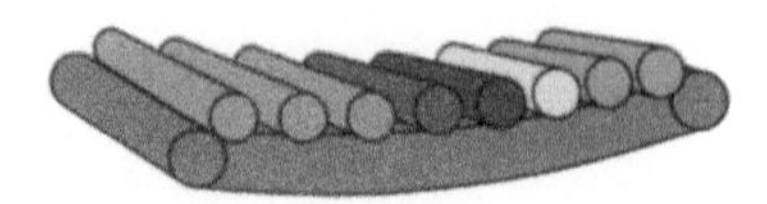

Figure 4.28. Cooling Bed.

Cooling Bed

Pipe is allowed to cool (see Figure 4.28).

Finishing

Straightening. Each pipe passes through a rotary straightener (see Figure 4.29).

Nondestructive Inspection. NDE is used to detect longitudinal and transverse flaws and ultrasonic testing is used to check the wall thickness (see Figure 4.30).

End Finishing. If required, bevelled ends are cut (see Figure 4.31).

Threading and Coupling. If required, the pipe is threaded and coupled at the mill (see Figure 4.32).

Hydrostatic Testing. All pipe lengths are then pressure tested to satisfy the relevant specification (see Figure 4.33).

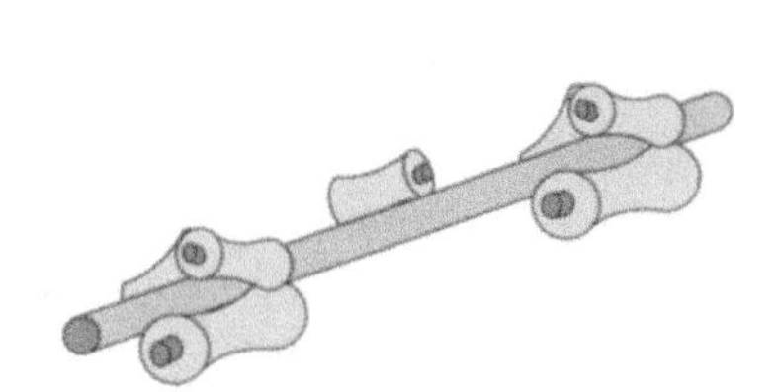

Figure 4.29. Rotary Straightener.

Figure 4.31. Beveller.

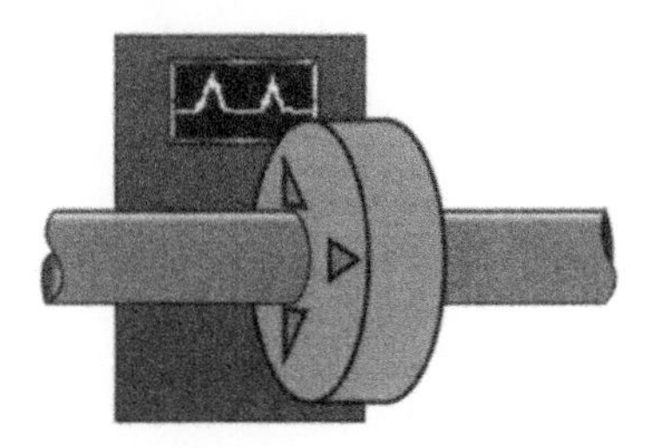

Figure 4.30. Nondestructive Examination.

Figure 4.32. Threading.

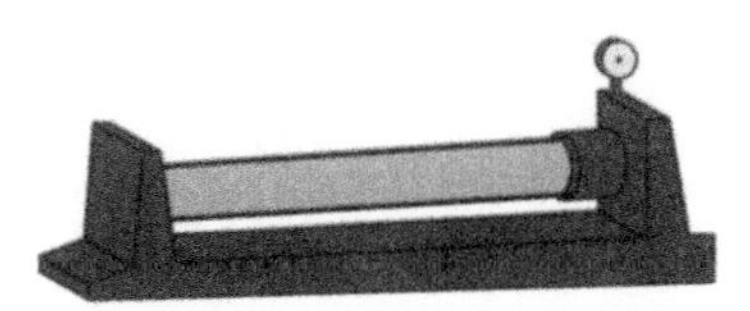

Figure 4.33. Pressure Testing.

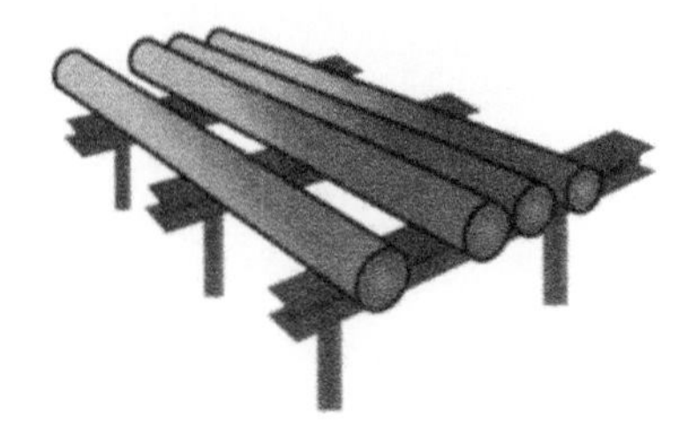

Figure 4.34. Pipes Awaiting Final Inspection.

Final Inspection. After a final visual inspection, the pipe is weighed, measured, and stenciled and either dispatched to the purchaser or held in stock (see Figure 4.34).

6. STEEL PIPE (SEAMLESS AND WELDED), DIMENSIONS AND WEIGHT

See Figure 4.35.

Dimensions—as per ASME B36.10.
Weights—as per manufacturer's estimates.
Size range—1/8 to 80 in., outside diameter (OD).
Wall thickness (WT)—STD, XS, XXS, and per schedules.
Methods of manufacture—covered in the relevant ASTM specification.
Chemical composition and mechanical properties—covered in the relevant ASTM specification.
Tolerances and permissible variations—depend on the method of manufacture, which are covered in the relevant ASTM specification.
Pipe length—covered in the relevant ASTM specification.
Below is a table that covers wall thicknesses and weights of pipe manufactured to ASME B36.19

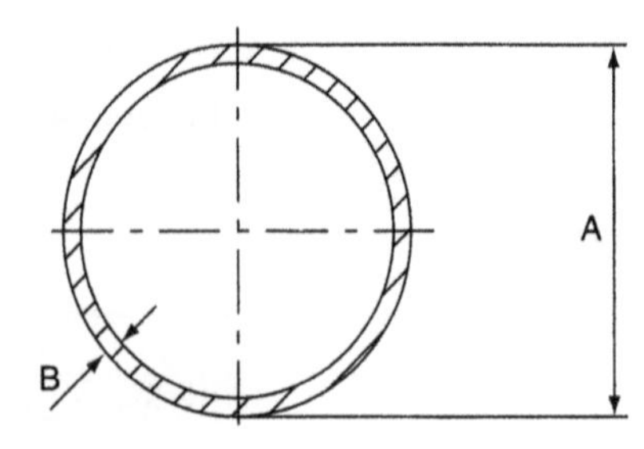

Figure 4.35. Taking the Dimensions of Steel Pipe.

NPS "A" in	DN "A" mm	**O.D. in**	O.D. mm	**wall.thk "B" in**	wall.thk "B" mm	**weight lb/ft**	weight kg/m	**Sch**	**Manf. Std.**
⅛"		**0.405**	10.3	**0.049**	1.24	**0.19**	0.28	10	
⅛"		**0.405**	10.3	**0.057**	1.45	**0.21**	0.32	30	
⅛"		**0.405**	10.3	**0.068**	1.73	**0.24**	0.37	40	STD
⅛"		**0.405**	10.3	**0.095**	2.41	**0.31**	0.47	80	XS
¼"		**0.540**	13.7	**0.065**	1.65	**0.33**	0.49	10	
¼"		**0.540**	13.7	**0.073**	1.85	**0.36**	0.54	30	
¼"		**0.540**	13.7	**0.088**	2.24	**0.43**	0.63	40	STD
¼"		**0.540**	13.7	**0.119**	3.02	**0.54**	0.80	80	XS
⅜"	10	**0.675**	17.1	**0.065**	1.65	**0.42**	0.63	10	
⅜"	10	**0.675**	17.1	**0.073**	1.85	**0.47**	0.70	30	
⅜"	10	**0.675**	17.1	**0.091**	2.31	**0.57**	0.84	40	STD
⅜"	10	**0.675**	17.1	**0.126**	3.20	**0.74**	1.10	80	XS
½"	15	**0.840**	21.3	**0.065**	1.65	**0.54**	0.80	5	
½"	15	**0.840**	21.3	**0.083**	2.11	**0.67**	1.00	10	
½"	15	**0.840**	21.3	**0.095**	2.41	**0.76**	1.12	30	
½"	15	**0.840**	21.3	**0.109**	2.77	**0.85**	1.27	40	STD
½"	15	**0.840**	21.3	**0.147**	3.73	**1.09**	1.62	80	XS
½"	15	**0.840**	21.3	**0.188**	4.78	**1.31**	1.95	160	
½"	15	**0.840**	21.3	**0.294**	7.47	**1.72**	2.55		XXS
¾"	20	**1.050**	26.7	**0.065**	1.65	**0.69**	1.03	5	
¾"	20	**1.050**	26.7	**0.083**	2.11	**0.86**	1.28	10	
¾"	20	**1.050**	26.7	**0.095**	2.41	**0.97**	1.44	30	
¾"	20	**1.050**	26.7	**0.113**	2.87	**1.13**	1.69	40	STD
¾"	20	**1.050**	26.7	**0.154**	3.91	**1.48**	2.20	80	XS
¾"	20	**1.050**	26.7	**0.219**	5.56	**1.95**	2.90	160	
¾"	20	**1.050**	26.7	**0.308**	7.82	**2.44**	3.64		XXS
1"	25	**1.315**	33.4	**0.065**	1.65	**0.87**	1.29	5	
1"	25	**1.315**	33.4	**0.109**	2.77	**1.41**	2.09	10	
1"	25	**1.315**	33.4	**0.114**	2.90	**1.46**	2.18	30	
1"	25	**1.315**	33.4	**0.133**	3.38	**1.68**	2.50	40	STD
1"	25	**1.315**	33.4	**0.179**	4.55	**2.17**	3.24	80	XS
1"	25	**1.315**	33.4	**0.250**	6.35	**2.85**	4.24	160	
1"	25	**1.315**	33.4	**0.358**	9.09	**3.66**	5.45		XXS
1 ¼"	32	**1.660**	42.2	**0.065**	1.65	**1.11**	1.65	5	
1 ¼"	32	**1.660**	42.2	**0.109**	2.77	**1.81**	2.69	10	
1 ¼"	32	**1.660**	42.2	**0.117**	2.97	**1.93**	2.87	30	
1 ¼"	32	**1.660**	42.2	**0.140**	3.56	**2.27**	3.39	40	STD

NPS "A" in	DN "A" mm	O.D. in	O.D. mm	wall.thk "B" in	wall.thk "B" mm	weight lb/ft	weight kg/m	Sch	Manf. Std.
1 ¼″	32	**1.660**	42.2	**0.191**	4.85	**3.00**	4.47	80	XS
1 ¼″	32	**1.660**	42.2	**0.250**	6.35	**3.77**	5.61	160	
1 ¼″	32	**1.660**	42.2	**0.382**	9.70	**5.22**	7.77		XXS
1 ½″	40	**1.900**	48.3	**0.065**	1.65	**1.28**	1.90	5	
1 ½″	40	**1.900**	48.3	**0.109**	2.77	**2.09**	3.11	10	
1 ½″	40	**1.900**	48.3	**0.125**	3.18	**2.37**	3.53	30	
1 ½″	40	**1.900**	48.3	**0.145**	3.68	**2.72**	4.05	40	STD
1 ½″	40	**1.900**	48.3	**0.200**	5.08	**3.63**	5.41	80	XS
1 ½″	40	**1.900**	48.3	**0.281**	7.14	**4.86**	7.25	160	
1 ½″	40	**1.900**	48.3	**0.400**	10.15	**6.41**	9.55		XXS
2″	50	**2.375**	60.3	**0.065**	1.65	**1.61**	2.39	5	
2″	50	**2.375**	60.3	**0.083**	2.11	**2.03**	3.03		
2″	50	**2.375**	60.3	**0.109**	2.77	**2.64**	3.93	10	
2″	50	**2.375**	60.3	**0.125**	3.18	**3.01**	4.48	30	
2″	50	**2.375**	60.3	**0.141**	3.58	**3.37**	5.01		
2″	50	**2.375**	60.3	**0.154**	3.91	**3.66**	5.44	40	STD
2″	50	**2.375**	60.3	**0.172**	4.37	**4.05**	6.03		
2″	50	**2.375**	60.3	**0.188**	4.78	**4.40**	6.54		
2″	50	**2.375**	60.3	**0.218**	5.54	**5.03**	7.48	80	XS
2″	50	**2.375**	60.3	**0.250**	6.35	**5.68**	8.45		
2″	50	**2.375**	60.3	**0.281**	7.14	**6.29**	9.36		
2″	50	**2.375**	60.3	**0.344**	8.74	**7.47**	11.11	160	
2″	50	**2.375**	60.3	**0.436**	11.07	**9.04**	13.44		XXS
2 ½″	65	**2.875**	73.0	**0.083**	2.11	**2.48**	3.69	5	
2 ½″	65	**2.875**	73.0	**0.109**	2.77	**3.22**	4.80		
2 ½″	65	**2.875**	73.0	**0.120**	3.05	**3.53**	5.25	10	
2 ½″	65	**2.875**	73.0	**0.125**	3.18	**3.67**	5.48		
2 ½″	65	**2.875**	73.0	**0.141**	3.58	**4.12**	6.13		
2 ½″	65	**2.875**	73.0	**0.156**	3.96	**4.53**	6.74		
2 ½″	65	**2.875**	73.0	**0.172**	4.37	**4.97**	7.40		
2 ½″	65	**2.875**	73.0	**0.188**	4.78	**5.40**	8.04	30	
2 ½″	65	**2.875**	73.0	**0.203**	5.16	**5.80**	8.63	40	STD
2 ½″	65	**2.875**	73.0	**0.216**	5.49	**6.14**	9.14		
2 ½″	65	**2.875**	73.0	**0.250**	6.35	**7.02**	10.44		
2 ½″	65	**2.875**	73.0	**0.276**	7.01	**7.67**	11.41	80	XS
2 ½″	65	**2.875**	73.0	**0.375**	9.53	**10.02**	14.92	160	
2 ½″	65	**2.875**	73.0	**0.552**	14.02	**13.71**	20.39		XXS

NPS "A" in	DN "A" mm	**O.D. in**	O.D. mm	**wall.thk "B" in**	wall.thk "B" mm	**weight lb/ft**	weight kg/m	**Sch**	**Manf. Std.**
3″	75	**3.500**	0.083	**0.083**	2.11	**3.03**	4.52	5	
3″	75	**3.500**	0.083	**0.109**	2.77	**3.95**	5.88		
3″	75	**3.500**	0.083	**0.102**	3.05	**4.34**	6.46	10	
3″	75	**3.500**	0.083	**0.125**	3.18	**4.51**	6.72		
3″	75	**3.500**	0.083	**0.141**	3.58	**5.06**	7.53		
3″	75	**3.500**	0.083	**0.156**	3.96	**5.58**	8.30		
3″	75	**3.500**	0.083	**0.172**	4.37	**6.12**	9.11		
3″	75	**3.500**	0.083	**0.188**	4.78	**6.66**	9.92	30	
3″	75	**3.500**	0.083	**0.216**	5.49	**7.58**	11.29	40	STD
3″	75	**3.500**	0.083	**0.250**	6.35	**8.69**	12.93		
3″	75	**3.500**	0.083	**0.281**	7.14	**9.67**	14.40		
3″	75	**3.500**	0.083	**0.300**	7.62	**10.26**	15.27	80	XS
3″	75	**3.500**	0.083	**0.438**	11.13	**14.34**	21.35	160	
3″	75	**3.500**	0.083	**0.600**	15.24	**18.60**	27.68		XXS
3 ½″		**4.000**	101.6	**0.083**	2.11	**3.48**	5.18	5	
3 ½″		**4.000**	101.6	**0.109**	2.77	**4.53**	6.75		
3 ½″		**4.000**	101.6	**0.120**	3.05	**4.98**	7.41	10	
3 ½″		**4.000**	101.6	**0.125**	3.18	**5.18**	7.72		
3 ½″		**4.000**	101.6	**0.141**	3.58	**5.82**	8.65		
3 ½″		**4.000**	101.6	**0.156**	3.96	**6.41**	9.54		
3 ½″		**4.000**	101.6	**0.172**	4.37	**7.04**	10.48		
3 ½″		**4.000**	101.6	**0.188**	4.78	**7.66**	11.41	30	
3 ½″		**4.000**	101.6	**0.226**	5.74	**9.12**	13.57	40	STD
3 ½″		**4.000**	101.6	**0.250**	6.35	**10.02**	14.92		
3 ½″		**4.000**	101.6	**0.281**	7.14	**11.17**	16.63		
3 ½″		**4.000**	101.6	**0.318**	8.08	**12.52**	18.64	80	XS
4″	100	**4.500**	114.3	**0.083**	2.11	**3.92**	5.84	5	
4″	100	**4.500**	114.3	**0.109**	2.77	**5.12**	7.62		
4″	100	**4.500**	114.3	**0.120**	3.05	**5.62**	8.37	10	
4″	100	**4.500**	114.3	**0.125**	3.18	**5.85**	8.71		
4″	100	**4.500**	114.3	**0.141**	3.58	**6.57**	9.78		
4″	100	**4.500**	114.3	**0.156**	3.96	**7.24**	10.78		
4″	100	**4.500**	114.3	**0.172**	4.37	**7.95**	11.85		
4″	100	**4.500**	114.3	**0.188**	4.78	**8.67**	12.91	30	
4″	100	**4.500**	114.3	**0.203**	5.16	**9.32**	13.89		
4″	100	**4.500**	114.3	**0.219**	5.56	**10.02**	14.91		
4″	100	**4.500**	114.3	**0.237**	6.02	**10.80**	16.08	40	STD
4″	100	**4.500**	114.3	**0.250**	6.35	**11.36**	16.91		

NPS "A" in	DN "A" mm	O.D. in	O.D. mm	wall.thk "B" in	wall.thk "B" mm	weight lb/ft	weight kg/m	Sch	Manf. Std.
4″	100	**4.500**	114.3	**0.281**	7.14	**12.67**	18.87		
4″	100	**4.500**	114.3	**0.312**	7.92	**13.97**	20.78		
4″	100	**4.500**	114.3	**0.337**	8.56	**15.00**	22.32	80	XS
4″	100	**4.500**	114.3	**0.438**	11.13	**19.02**	28.32	120	
4″	100	**4.500**	114.3	**0.531**	13.49	**22.53**	33.54	160	
4″	100	**4.500**	114.3	**0.674**	17.12	**27.57**	41.03		XXS
5″	125	**5.563**	141.3	**0.083**	2.11	**4.86**	7.24		
5″	125	**5.563**	141.3	**0.119**	2.77	**6.36**	9.46	5	
5″	125	**5.563**	141.3	**0.125**	3.18	**7.27**	10.83		
5″	125	**5.563**	141.3	**0.134**	3.40	**7.78**	11.56	10	
5″	125	**5.563**	141.3	**0.156**	3.96	**9.02**	13.41		
5″	125	**5.563**	141.3	**0.188**	4.78	**10.80**	16.09		
5″	125	**5.563**	141.3	**0.219**	5.56	**12.51**	18.61		
5″	125	**5.563**	141.3	**0.258**	6.55	**14.63**	21.77	40	STD
5″	125	**5.563**	141.3	**0.281**	7.14	**15.87**	23.62		
5″	125	**5.563**	141.3	**0.312**	7.92	**17.51**	26.05		
5″	125	**5.563**	141.3	**0.344**	8.74	**19.19**	28.57		
5″	125	**5.563**	141.3	**0.375**	9.53	**20.80**	30.97	80	XS
5″	125	**5.563**	141.3	**0.500**	12.70	**27.06**	40.28	120	
5″	125	**5.563**	141.3	**0.625**	15.88	**32.99**	49.12	160	
5″	125	**5.563**	141.3	**0.750**	19.05	**38.58**	57.43		XXS
6″	150	**6.625**	168.3	**0.083**	2.11	**5.80**	8.65		
6″	150	**6.625**	168.3	**0.109**	2.77	**7.59**	11.31	5	
6″	150	**6.625**	168.3	**0.125**	3.18	**8.69**	12.95		
6″	150	**6.625**	168.3	**0.134**	3.40	**9.30**	13.83	10	
6″	150	**6.625**	168.3	**0.141**	3.58	**9.77**	14.54		
6″	150	**6.625**	168.3	**0.156**	3.96	**10.79**	16.05		
6″	150	**6.625**	168.3	**0.172**	4.37	**11.87**	17.67		
6″	150	**6.625**	168.3	**0.188**	4.78	**12.94**	19.28		
6″	150	**6.625**	168.3	**0.203**	5.16	**13.94**	20.76		
6″	150	**6.625**	168.3	**0.219**	5.56	**15.00**	22.31		
6″	150	**6.625**	168.3	**0.250**	6.35	**17.04**	25.36		
6″	150	**6.625**	168.3	**0.280**	7.11	**18.99**	28.26	40	STD
6″	150	**6.625**	168.3	**0.312**	7.92	**21.06**	31.33		
6″	150	**6.625**	168.3	**0.344**	8.74	**23.10**	34.39		
6″	150	**6.625**	168.3	**0.375**	9.53	**25.05**	37.31		
6″	150	**6.625**	168.3	**0.432**	10.67	**28.60**	42.56	80	XS

NPS "A" in	DN "A" mm	**O.D. in**	O.D. mm	**wall.thk "B" in**	wall.thk "B" mm	**weight lb/ft**	weight kg/m	**Sch**	**Manf. Std.**
6″	150	**6.625**	168.3	**0.500**	12.70	**32.74**	48.73		
6″	150	**6.625**	168.3	**0.562**	14.27	**36.43**	54.21	120	
6″	150	**6.625**	168.3	**0.625**	15.88	**40.09**	59.69		
6″	150	**6.625**	168.3	**0.719**	18.26	**45.39**	67.57	160	
6″	150	**6.625**	168.3	**0.750**	19.05	**47.10**	70.12		
6″	150	**6.625**	168.3	**0.864**	21.95	**53.21**	79.22		XXS
6″	150	**6.625**	168.3	**0.875**	22.23	**53.78**	80.08		
8″	200	**8.625**	219.1	**0.109**	2.77	**9.92**	14.78	5	
8″	200	**8.625**	219.1	**0.125**	3.18	**11.36**	16.93		
8″	200	**8.625**	219.1	**0.148**	3.76	**13.41**	19.97	10	
8″	200	**8.625**	219.1	**0.158**	3.96	**14.12**	21.01		
8″	200	**8.625**	219.1	**0.188**	4.78	**16.96**	25.26		
8″	200	**8.625**	219.1	**0.203**	5.16	**18.28**	27.22		
8″	200	**8.625**	219.1	**0.219**	5.56	**19.68**	29.28		
8″	200	**8.625**	219.1	**0.250**	6.35	**22.38**	33.32	20	
8″	200	**8.625**	219.1	**0.277**	7.04	**24.72**	36.82	30	
8″	200	**8.625**	219.1	**0.312**	7.92	**27.58**	41.25		
8″	200	**8.625**	219.1	**0.322**	8.18	**28.58**	42.55	40	STD
8″	200	**8.625**	219.1	**0.344**	8.74	**30.45**	45.34		
8″	200	**8.625**	219.1	**0.375**	9.53	**33.07**	49.25		
8″	200	**8.625**	219.1	**0.406**	10.31	**35.67**	53.09	60	
8″	200	**8.625**	219.1	**0.438**	11.13	**38.33**	57.08		
8″	200	**8.625**	219.1	**0.500**	12.70	**43.43**	64.64	80	XS
8″	200	**8.625**	219.1	**0.562**	14.27	**48.44**	72.08		
8″	200	**8.625**	219.1	**0.594**	15.09	**51.00**	75.92	100	
8″	200	**8.625**	219.1	**0.625**	15.88	**53.45**	79.59		
8″	200	**8.625**	219.1	**0.719**	18.26	**60.77**	90.44	120	
8″	200	**8.625**	219.1	**0.750**	19.05	**63.14**	93.98		
8″	200	**8.625**	219.1	**0.812**	20.62	**67.82**	100.93	140	
8″	200	**8.625**	219.1	**0.875**	20.23	**72.49**	107.93		XXS
8″	200	**8.625**	219.1	**0.906**	23.01	**74.76**	111.27	160	
8″	200	**8.625**	219.1	**1.000**	25.40	**81.51**	121.33		
10″	250	**10.750**	273.0	**0.134**	3.40	**15.21**	22.61	5	
10″	250	**10.750**	273.0	**0.156**	3.96	**17.67**	26.27		
10″	250	**10.750**	273.0	**0.165**	4.19	**18.67**	27.78	10	
10″	250	**10.750**	273.0	**0.188**	4.78	**21.23**	31.62		

NPS "A" in	DN "A" mm	**O.D. in**	O.D. mm	**wall.thk "B" in**	wall.thk "B" mm	**weight lb/ft**	weight kg/m	**Sch**	**Manf. Std.**
10″	250	**10.750**	273.0	**0.209**	5.16	**22.89**	34.08		
10″	250	**10.750**	273.0	**0.219**	5.56	**24.65**	36.67		
10″	250	**10.750**	273.0	**0.250**	6.35	**28.06**	41.76	20	
10″	250	**10.750**	273.0	**0.279**	7.09	**31.23**	46.49		
10″	250	**10.750**	273.0	**0.307**	7.80	**34.27**	51.01	30	
10″	250	**10.750**	273.0	**0.344**	6.74	**38.27**	56.96		
10″	250	**10.750**	273.0	**0.365**	9.27	**40.52**	60.29	40	STD
10″	250	**10.750**	273.0	**0.438**	11.13	**48.28**	71.88		
10″	250	**10.750**	273.0	**0.500**	12.70	**54.79**	81.53	60	XS
10″	250	**10.750**	273.0	**0.562**	14.27	**61.21**	91.05		
10″	250	**10.750**	273.0	**0.594**	15.09	**64.49**	95.98		80
10″	250	**10.750**	273.0	**0.625**	15.88	**67.65**	100.69		
10″	250	**10.750**	273.0	**0.719**	18.26	**77.10**	114.71		100
10″	250	**10.750**	273.0	**0.812**	20.62	**86.26**	128.34		
10″	250	**10.750**	273.0	**0.844**	21.44	**89.38**	133.01		120
10″	250	**10.750**	273.0	**0.875**	22.23	**92.37**	137.48		
10″	250	**10.750**	273.0	**0.938**	23.83	**98.39**	146.43		
10″	250	**10.750**	273.0	**1.000**	25.40	**104.23**	155.10	XXS	140
10″	250	**10.750**	273.0	**1.125**	28.58	**115.75**	172.27		160
10″	250	**10.750**	273.0	**1.250**	31.75	**126.94**	188.90		
12″	300	**12.750**	323.8	**0.156**	3.96	**21.00**	31.24	5	
12″	300	**12.750**	323.8	**0.172**	4.37	**23.13**	34.43		
12″	300	**12.750**	323.8	**0.180**	4.57	**24.19**	35.98	10	
12″	300	**12.750**	323.8	**0.188**	4.78	**25.25**	37.61		
12″	300	**12.750**	323.8	**0.203**	5.16	**27.23**	40.55		
12″	300	**12.750**	323.8	**0.219**	5.56	**29.34**	43.64		
12″	300	**12.750**	323.8	**0.250**	6.35	**33.41**	49.71	20	
12″	300	**12.750**	323.8	**0.281**	7.14	**37.46**	55.76		
12″	300	**12.750**	323.8	**0.312**	7.92	**41.48**	61.70		
12″	300	**12.750**	323.8	**0.330**	8.38	**43.81**	65.19	30	
12″	300	**12.750**	323.8	**0.344**	8.74	**45.62**	67.91		
12″	300	**12.750**	323.8	**0.375**	9.53	**49.61**	73.86		STD
12″	300	**12.750**	323.8	**0.406**	10.31	**53.57**	79.71	40	
12″	300	**12.750**	323.8	**0.438**	11.13	**57.65**	85.82		
12″	300	**12.750**	323.8	**0.500**	12.70	**65.48**	97.44		
12″	300	**12.750**	323.8	**0.562**	14.27	**73.22**	108.93	60	

NPS "A" in	DN "A" mm	**O.D. in**	O.D. mm	**wall.thk "B" in**	wall.thk "B" mm	**weight lb/ft**	weight kg/m	**Sch**	**Manf. Std.**
12″	300	**12.750**	323.8	**0.625**	15.88	**81.01**	120.59		
12″	300	**12.750**	323.8	**0.688**	17.48	**88.71**	132.05	80	
12″	300	**12.750**	323.8	**0.750**	19.05	**96.21**	143.17		
12″	300	**12.750**	323.8	**0.812**	20.62	**103.63**	154.17		
12″	300	**12.750**	323.8	**0.844**	21.44	**107.42**	159.87	100	
12″	300	**12.750**	323.8	**0.875**	22.23	**111.08**	165.33		
12″	300	**12.750**	323.8	**0.938**	23.83	**118.44**	176.29		
12″	300	**12.750**	323.8	**1.000**	25.40	**125.61**	186.92	120	XXS
12″	300	**12.750**	323.8	**1.062**	26.97	**132.69**	197.43		
12″	300	**12.750**	323.8	**1.125**	28.58	**139.81**	208.08	140	
12″	300	**12.750**	323.8	**1.250**	31.75	**153.67**	228.68		
12″	300	**12.750**	323.8	**1.312**	33.32	**160.42**	238.69	160	
14″	350	**14.000**	355.6	**0.156**	3.96	**23.09**	34.34	5	
14″	350	**14.000**	355.6	**0.188**	4.78	**27.76**	41.36		
14″	350	**14.000**	355.6	**0.203**	5.16	**29.94**	44.59		
14″	350	**14.000**	355.6	**0.210**	5.33	**30.96**	46.04		
14″	350	**14.000**	355.6	**0.219**	5.56	**32.26**	48.00		
14″	350	**14.000**	355.6	**0.250**	6.35	**36.75**	54.69	10	
14″	350	**14.000**	355.6	**0.281**	7.14	**41.21**	61.36		
14″	350	**14.000**	355.6	**0.312**	7.92	**45.65**	67.91	20	
14″	350	**14.000**	355.6	**0.344**	8.74	**50.22**	74.76		
14″	350	**14.000**	355.6	**0.375**	9.53	**54.62**	81.33	30	STD
14″	350	**14.000**	355.6	**0.406**	10.31	**59.00**	87.79		
14″	350	**14.000**	355.6	**0.438**	11.13	**63.50**	94.55	40	
14″	350	**14.000**	355.6	**0.469**	11.91	**67.84**	100.95		
14″	350	**14.000**	355.6	**0.500**	12.70	**72.16**	107.40		XS
14″	350	**14.000**	355.6	**0.562**	14.27	**80.73**	120.12		
14″	350	**14.000**	355.6	**0.594**	15.09	**85.13**	126.72	60	
14″	350	**14.000**	355.6	**0.625**	15.88	**89.36**	133.04		
14″	350	**14.000**	355.6	**0.688**	17.48	**97.91**	145.76		
14″	350	**14.000**	355.6	**0.750**	19.05	**106.23**	158.11	80	
14″	350	**14.000**	355.6	**0.812**	20.62	**114.48**	170.34		
14″	350	**14.000**	355.6	**0.875**	22.32	**122.77**	182.76		
14″	350	**14.000**	355.6	**0.938**	23.83	**130.98**	194.98	100	
14″	350	**14.000**	355.6	**1.000**	25.40	**138.97**	208.84		
14″	350	**14.000**	355.6	**1.062**	26.97	**146.88**	218.58		

NPS "A" in	DN "A" mm	O.D. in	O.D. mm	wall.thk "B" in	wall.thk "B" mm	weight lb/ft	weight kg/m	Sch	Manf. Std.
14″	350	**14.000**	355.6	**1.094**	27.79	**150.93**	224.66	120	
14″	350	**14.000**	355.6	**1.125**	28.58	**154.84**	230.49		
14″	350	**14.000**	355.6	**1.250**	31.75	**170.37**	253.58	140	
14″	350	**14.000**	355.6	**1.406**	35.71	**189.29**	281.72	160	
14″	350	**14.000**	355.6	**2.000**	50.80	**256.56**	381.85		
14″	350	**14.000**	355.6	**2.125**	53.98	**269.76**	401.52		
14″	350	**14.000**	355.6	**2.200**	55.88	**277.51**	413.04		
14″	350	**14.000**	355.6	**2.500**	63.50	**307.34**	457.43		
16″	400	**16.00**	406.4	**0.165**	4.19	**27.93**	41.56	5	
16″	400	**16.00**	406.4	**0.188**	4.78	**31.78**	47.34		
16″	400	**16.00**	406.4	**0.203**	5.16	**34.28**	51.06		
16″	400	**16.00**	406.4	**0.219**	5.56	**36.95**	54.96		
16″	400	**16.00**	406.4	**0.250**	6.35	**42.09**	62.65	10	
16″	400	**16.00**	406.4	**0.281**	7.14	**47.22**	70.30		
16″	400	**16.00**	406.4	**0.312**	7.92	**52.32**	77.83	20	
16″	400	**16.00**	406.4	**0.344**	8.74	**57.57**	85.71		
16″	400	**16.00**	406.4	**0.375**	9.53	**62.64**	93.27	30	STD
16″	400	**16.00**	406.4	**0.406**	10.31	**67.68**	100.71		
16″	400	**16.00**	406.4	**0.438**	11.13	**72.86**	108.49		
16″	400	**16.00**	406.4	**0.469**	11.91	**77.87**	115.87		
16″	400	**16.00**	406.4	**0.500**	12.70	**82.85**	123.31	40	XS
16″	400	**16.00**	406.4	**0.562**	14.27	**92.75**	138.00		
16″	400	**16.00**	406.4	**0.625**	15.88	**102.72**	152.94		
16″	400	**16.00**	406.4	**0.656**	16.66	**107.60**	160.13	60	
16″	400	**16.00**	406.4	**0.688**	17.48	**112.62**	167.66		
16″	400	**16.00**	406.4	**0.750**	19.05	**122.27**	181.98		
16″	400	**16.00**	406.4	**0.812**	20.62	**131.84**	196.18		
16″	400	**16.00**	406.4	**0.844**	21.44	**136.74**	203.54	80	
16″	400	**16.00**	406.4	**0.875**	22.23	**141.48**	210.61		
16″	400	**16.00**	406.4	**0.938**	23.83	**151.03**	224.83		
16″	400	**16.00**	406.4	**1.000**	25.40	**160.35**	238.66		
16″	400	**16.00**	406.4	**1.031**	26.19	**164.98**	245.57	100	
16″	400	**16.00**	406.4	**1.062**	26.97	**169.59**	252.37		
16″	400	**16.00**	406.4	**1.125**	28.58	**178.89**	266.30		
16″	400	**16.00**	406.4	**1.188**	30.18	**188.11**	280.01		
16″	400	**16.00**	406.4	**1.219**	30.96	**193.61**	286.66	120	
16″	400	**16.00**	406.4	**1.250**	31.75	**197.10**	293.35		
16″	400	**16.00**	406.4	**1.438**	36.53	**223.85**	333.21	140	
16″	400	**16.00**	406.4	**1.594**	40.49	**245.48**	365.38	160	

NPS "A" in	DN "A" mm	**O.D. in**	O.D. mm	**wall.thk "B" in**	wall.thk "B" mm	**weight lb/ft**	weight kg/m	**Sch**	**Manf. Std.**
18″	450	**18.00**	457.0	**0.165**	4.19	**31.46**	46.79	5	
18″	450	**18.00**	457.0	**0.188**	4.78	**35.80**	53.31		
18″	450	**18.00**	457.0	**0.219**	5.56	**41.63**	61.90		
18″	450	**18.00**	457.0	**0.250**	6.35	**47.44**	70.57	10	
18″	450	**18.00**	457.0	**0.281**	7.14	**53.23**	79.21		
18″	450	**18.00**	457.0	**0.312**	7.92	**58.99**	87.71	20	
18″	450	**18.00**	457.0	**0.344**	8.74	**64.93**	96.62		
18″	450	**18.00**	457.0	**0.375**	9.53	**70.65**	105.17		STD
18″	450	**18.00**	457.0	**0.406**	10.31	**76.36**	113.58		
18″	450	**18.00**	457.0	**0.438**	11.13	**82.23**	122.38	30	
18″	450	**18.00**	457.0	**0.469**	11.91	**87.89**	130.73		
18″	450	**18.00**	457.0	**0.500**	12.70	**93.54**	139.16		XS
18″	450	**18.00**	457.0	**0.562**	14.27	**104.76**	155.81	40	
18″	450	**18.00**	457.0	**0.625**	15.88	**116.09**	172.75		
18″	450	**18.00**	457.0	**0.688**	17.48	**127.32**	189.47		
18″	450	**18.00**	457.0	**0.750**	19.05	**138.30**	205.75	60	
18″	450	**18.00**	457.0	**0.812**	20.62	**149.20**	221.91		
18″	450	**18.00**	457.0	**0.875**	22.23	**160.20**	238.35		
18″	450	**18.00**	457.0	**0.938**	23.83	**171.08**	254.57	80	
18″	450	**18.00**	457.0	**1.000**	25.40	**181.73**	270.36		
18″	450	**18.00**	457.0	**1.062**	26.97	**192.29**	286.02		
18″	450	**18.00**	457.0	**1.125**	28.58	**202.94**	301.96		
18″	450	**18.00**	457.0	**1.156**	29.36	**208.15**	309.64	100	
18″	450	**18.00**	457.0	**1.188**	30.18	**213.51**	317.68		
18″	450	**18.00**	457.0	**1.250**	31.75	**223.82**	332.97		
18″	450	**18.00**	457.0	**1.375**	34.93	**244.37**	363.58	120	
18″	450	**18.00**	457.0	**1.562**	39.67	**274.48**	408.28	140	
18″	450	**18.00**	457.0	**1.781**	45.24	**308.79**	459.39	160	
20″	500	**20.00**	508	**0.188**	4.78	**39.82**	59.32	5	
20″	500	**20.00**	508	**0.219**	5.56	**46.31**	68.89		
20″	500	**20.00**	508	**0.250**	6.35	**52.78**	78.56	10	
20″	500	**20.00**	508	**0.281**	7.14	**59.23**	88.19		
20″	500	**20.00**	508	**0.312**	7.92	**65.66**	97.68		
20″	500	**20.00**	508	**0.344**	8.74	**72.28**	107.61		
20″	500	**20.00**	508	**0.375**	9.53	**78.67**	117.15	20	STD
20″	500	**20.00**	508	**0.406**	10.31	**85.04**	126.54		

NPS "A" in	DN "A" mm	O.D. in	O.D. mm	wall.thk "B" in	wall.thk "B" mm	weight lb/ft	weight kg/m	Sch	Manf. Std.
20″	500	20.00	508	0.438	11.13	91.59	136.38		
20″	500	20.00	508	0.469	11.91	97.92	145.71		
20″	500	20.00	508	0.500	12.70	104.23	155.13	30	XS
20″	500	20.00	508	0.562	14.27	116.78	173.75		
20″	500	20.00	508	0.594	15.09	123.23	183.43	40	
20″	500	20.00	508	0.625	15.88	129.45	192.73		
20″	500	20.00	508	0.688	17.48	142.03	211.45		
20″	500	20.00	508	0.750	19.05	154.34	229.71		
20″	500	20.00	508	0.812	20.62	166.56	247.84	60	
20″	500	20.00	508	0.875	22.23	178.89	266.31		
20″	500	20.00	508	0.938	23.83	191.14	284.54		
20″	500	20.00	508	1.000	25.40	203.11	302.30		
20″	500	20.00	508	1.031	26.19	209.06	311.19	80	
20″	500	20.00	508	1.062	26.97	215.00	319.94		
20″	500	20.00	508	1.125	28.58	227.00	337.91		
20″	500	20.00	508	1.188	30.18	238.91	355.63		
20″	500	20.00	508	1.250	31.75	250.55	372.91		
20″	500	20.00	508	1.281	32.54	256.34	381.55	100	
20″	500	20.00	508	1.312	33.32	262.10	390.05		
20″	500	20.00	508	1.375	34.93	273.76	407.51		
20″	500	20.00	508	1.500	38.10	296.65	441.52	120	
20″	500	20.00	508	1.750	44.45	341.41	508.15	140	
20″	500	20.00	508	1.969	50.1	378.53	564.85	160	
22″	550	22.00	559	0.188	4.78	43.84	65.33	5	
22″	550	22.00	559	0.219	5.56	50.99	75.89		
22″	550	22.00	559	0.250	6.35	58.13	86.55	10	
22″	550	22.00	559	0.281	7.14	65.24	97.17		
22″	550	22.00	559	0.312	7.92	72.34	107.84		
22″	550	22.00	559	0.344	8.74	79.64	118.60		
22″	550	22.00	559	0.375	9.53	86.69	129.14	20	STD
22″	550	22.00	559	0.406	10.31	93.72	139.51		
22″	550	22.00	559	0.438	11.13	100.96	150.38		
22″	550	22.00	559	0.469	11.91	107.95	160.69		
22″	550	22.00	559	0.500	12.70	114.92	171.10	30	XS
22″	550	22.00	559	0.562	14.27	128.79	191.70		

NPS "A" in	DN "A" mm	O.D. in	O.D. mm	wall.thk "B" in	wall.thk "B" mm	weight lb/ft	weight kg/m	Sch	Manf. Std.
22″	550	22.00	559	0.625	15.88	142.81	212.70		
22″	550	22.00	559	0.688	17.48	156.74	233.44		
22″	550	22.00	559	0.750	19.05	170.37	253.67		
22″	550	22.00	559	0.812	20.62	183.92	273.78		
22″	550	22.00	559	0.875	22.23	197.60	294.27	60	
22″	550	22.00	559	0.938	23.83	211.19	314.51		
22″	550	22.00	559	1.000	25.40	224.49	334.25		
22″	550	22.00	559	1.062	26.97	237.70	353.86		
22″	550	22.00	559	1.125	28.58	251.05	373.85	80	
22″	550	22.00	559	1.188	30.18	264.31	393.59		
22″	550	22.00	559	1.250	31.75	277.27	412.84		
22″	550	22.00	559	1.312	33.32	290.15	431.96		
22″	550	22.00	559	1.375	34.93	303.16	451.45	100	
22″	550	22.00	559	1.438	36.53	316.08	470.69		
22″	550	22.00	559	1.500	38.10	328.72	489.44		
22″	550	22.00	559	1.625	41.28	353.94	527.05	120	
22″	550	22.00	559	1.875	47.63	403.38	600.30	140	
22″	550	22.00	559	2.125	53.98	451.49	672.30	160	
24″	600	24.00	610	0.218	5.54	55.42	82.58	5	
24″	600	24.00	610	0.250	6.35	63.47	94.53	10	
24″	600	24.00	610	0.281	7.14	71.25	106.15		
24″	600	24.00	610	0.312	7.92	79.01	117.60		
24″	600	24.00	610	0.344	8.74	86.99	129.60		
24″	600	24.00	610	0.375	9.53	94.71	141.12	20	STD
24″	600	24.00	610	0.406	10.31	102.40	152.48		
24″	600	24.00	610	0.438	11.13	110.32	164.38		
24″	600	24.00	610	0.469	11.91	117.98	175.67		
24″	600	24.00	610	0.500	12.70	125.61	187.07		XS
24″	600	24.00	610	0.562	14.27	140.81	209.65	30	
24″	600	24.00	610	0.625	15.88	156.17	232.67		
24″	600	24.00	610	0.688	17.48	171.45	255.43	40	
24″	600	24.00	610	0.750	19.05	186.41	277.63		
24″	600	24.00	610	0.812	20.62	201.28	299.71		
24″	600	24.00	610	0.875	22.23	216.31	322.23		
24″	600	24.00	610	0.938	23.83	231.25	344.48		
24″	600	24.00	610	0.969	24.61	238.57	355.28	60	
24″	600	24.00	610	1.000	25.40	245.87	366.19		
24″	600	24.00	610	1.062	26.97	260.41	387.79		

NPS "A" in	DN "A" mm	O.D. in	O.D. mm	wall.thk "B" in	wall.thk "B" mm	weight lb/ft	weight kg/m	Sch	Manf. Std.
24″	600	**24.00**	610	**1.125**	28.58	**275.10**	409.80		
24″	600	**24.00**	610	**1.188**	30.18	**289.71**	431.55		
24″	600	**24.00**	610	**1.219**	30.96	**296.86**	442.11	80	
24″	600	**24.00**	610	**1.250**	31.75	**304.00**	452.77		
24″	600	**24.00**	610	**1.312**	33.32	**318.21**	473.87		
24″	600	**24.00**	610	**1.375**	34.93	**332.56**	495.38		
24″	600	**24.00**	610	**1.438**	36.53	**346.83**	516.63		
24″	600	**24.00**	610	**1.500**	38.10	**360.79**	537.36		
24″	600	**24.00**	610	**1.531**	38.89	**367.74**	547.74	100	
24″	600	**24.00**	610	**1.562**	39.67	**374.66**	557.97		
24″	600	**24.00**	610	**1.812**	46.02	**429.79**	640.07	120	
24″	600	**24.00**	610	**2.062**	52.37	**483.57**	720.19	140	
24″	600	**24.00**	610	**2.344**	59.54	**542.64**	808.27	160	
26″	650	**26.00**	660	**0.250**	6.35	**68.82**	102.36		
26″	650	**26.00**	660	**0.281**	7.14	**77.26**	114.96		
26″	650	**26.00**	660	**0.312**	7.92	**85.68**	127.36	10	
26″	650	**26.00**	660	**0.344**	8.74	**94.35**	140.37		
26″	650	**26.00**	660	**0.375**	9.53	**102.72**	152.88		STD
26″	650	**26.00**	660	**0.405**	10.31	**111.08**	165.19		
26″	650	**26.00**	660	**0.438**	11.13	**119.69**	178.10		
26″	650	**26.00**	660	**0.469**	11.91	**128.00**	190.36		
26″	650	**26.00**	660	**0.500**	12.70	**136.30**	202.74	20	XS
26″	650	**26.00**	660	**0.562**	14.27	**152.83**	227.25		
26″	650	**26.00**	660	**0.625**	15.88	**169.54**	252.25		
26″	650	**26.00**	660	**0.688**	17.48	**186.16**	276.98		
26″	650	**26.00**	660	**0.750**	19.05	**202.44**	301.12		
26″	650	**26.00**	660	**0.812**	20.62	**218.64**	325.14		
26″	650	**26.00**	660	**0.875**	22.23	**235.01**	349.64		
26″	650	**26.00**	660	**0.938**	23.83	**251.30**	373.87		
26″	650	**26.00**	660	**1.000**	25.40	**267.25**	397.51		
28″	700	**28.00**	711	**0.250**	6.35	**74.16**	110.35		
28″	700	**28.00**	711	**0.281**	7.14	**83.26**	123.94		
28″	700	**28.00**	711	**0.312**	7.92	**92.35**	137.32	10	
28″	700	**28.00**	711	**0.344**	8.74	**101.70**	151.37		
28″	700	**28.00**	711	**0.375**	9.53	**110.74**	164.86		STD
28″	700	**28.00**	711	**0.406**	10.31	**119.76**	178.16		
28″	700	**28.00**	711	**0.438**	11.13	**129.05**	192.10		
28″	700	**28.00**	711	**0.469**	11.91	**138.03**	205.34		

NPS "A" in	DN "A" mm	O.D. in	O.D. mm	wall.thk "B" in	wall.thk "B" mm	weight lb/ft	weight kg/m	Sch	Manf. Std.
28″	700	**28.00**	711	**0.500**	12.70	**146.99**	218.71	20	XS
28″	700	**28.00**	711	**0.562**	14.27	**164.64**	245.19		
28″	700	**28.00**	711	**0.625**	15.88	**182.90**	272.23	30	
28″	700	**28.00**	711	**0.688**	17.48	**200.87**	298.96		
28″	700	**28.00**	711	**0.750**	19.05	**218.48**	325.08		
28″	700	**28.00**	711	**0.812**	20.62	**236.00**	351.07		
28″	700	**28.00**	711	**0.875**	22.23	**253.72**	377.60		
28″	700	**28.00**	711	**0.938**	23.83	**271.36**	403.84		
28″	700	**28.00**	711	**1.000**	25.40	**288.63**	429.46		
30″	750	**30.00**	762	**0.250**	6.35	**79.51**	118.34	5	
30″	750	**30.00**	762	**0.281**	7.14	**89.27**	134.92		
30″	750	**30.00**	762	**0.312**	7.92	**99.02**	147.29	10	
30″	750	**30.00**	762	**0.344**	8.74	**109.06**	162.36		
30″	750	**30.00**	762	**0.375**	9.53	**118.76**	176.85		STD
30″	750	**30.00**	762	**0.406**	10.31	**128.44**	191.12		
30″	750	**30.00**	762	**0.438**	11.13	**138.42**	206.10		
30″	750	**30.00**	762	**0.469**	11.91	**148.06**	220.32		
30″	750	**30.00**	762	**0.500**	12.70	**157.68**	234.68	20	XS
30″	750	**30.00**	762	**0.562**	14.27	**176.86**	263.14		
30″	750	**30.00**	762	**0.625**	15.88	**196.26**	292.20	30	
30″	750	**30.00**	762	**0.688**	17.48	**215.58**	320.95		
30″	750	**30.00**	762	**0.750**	19.05	**234.51**	349.04		
30″	750	**30.00**	762	**0.812**	20.62	**253.36**	377.01		
30″	750	**30.00**	762	**0.875**	22.23	**272.43**	405.56		
30″	750	**30.00**	762	**0.938**	23.83	**291.41**	433.81		
30″	750	**30.00**	762	**1.000**	25.40	**310.01**	461.41		
30″	750	**30.00**	762	**1.062**	26.97	**328.53**	488.88		
30″	750	**30.00**	762	**1.125**	28.58	**347.26**	516.93		
30″	750	**30.00**	762	**1.188**	30.18	**365.90**	544.68		
30″	750	**30.00**	762	**1.250**	31.75	**384.17**	571.79		
32″	800	**32.00**	813	**0.250**	6.35	**84.85**	126.32		
32″	800	**32.00**	813	**0.281**	7.14	**95.28**	141.90		
32″	800	**32.00**	813	**0.312**	7.92	**105.69**	157.25	10	
32″	800	**32.00**	813	**0.344**	8.74	**116.41**	173.35		
32″	800	**32.00**	813	**0.375**	9.53	**126.78**	188.83		STD
32″	800	**32.00**	813	**0.406**	10.31	**137.12**	204.09		
32″	800	**32.00**	813	**0.438**	11.13	**147.78**	220.10		
32″	800	**32.00**	813	**0.469**	11.91	**158.08**	235.29		

NPS "A" in	DN "A" mm	O.D. in	O.D. mm	wall.thk "B" in	wall.thk "B" mm	weight lb/ft	weight kg/m	Sch	Manf. Std.
32″	800	32.00	813	0.500	12.70	168.37	250.65	20	XS
32″	800	32.00	813	0.562	14.27	188.87	281.09		
32″	800	32.00	813	0.625	15.88	209.62	312.17	30	
32″	800	32.00	813	0.688	17.48	230.29	342.94	40	
32″	800	32.00	813	0.750	19.05	250.55	373.00		
32″	800	32.00	813	0.812	20.62	270.72	402.94		
32″	800	32.00	813	0.875	22.23	291.14	433.52		
32″	800	32.00	813	0.938	23.83	311.47	463.78		
32″	800	32.00	813	1.000	25.40	331.39	493.35		
32″	800	32.00	813	1.062	26.97	351.23	522.80		
32″	800	32.00	813	1.125	28.58	371.31	552.88		
32″	800	32.00	813	1.188	30.18	391.30	582.64		
32″	800	32.00	813	1.250	31.75	410.90	611.72		
34″	850	34.00	864	0.250	6.35	90.20	134.31		
34″	850	34.00	864	0.281	7.14	101.29	150.88		
34″	850	34.00	864	0.312	7.92	112.36	167.21	10	
34″	850	34.00	864	0.344	8.74	123.77	184.34		
34″	850	34.00	864	0.375	9.53	134.79	200.82		STD
34″	850	34.00	864	0.406	10.31	145.80	217.06		
34″	850	34.00	864	0.438	11.13	157.14	234.10		
34″	850	34.00	864	0.469	11.91	168.11	250.27		
34″	850	34.00	864	0.500	12.70	179.06	266.63	20	XS
34″	850	34.00	864	0.562	14.27	200.89	299.04		
34″	850	34.00	864	0.625	15.88	222.99	332.14	30	
34″	850	34.00	864	0.688	17.48	245.00	364.92	40	
34″	850	34.00	864	0.750	19.05	266.58	396.96		
34″	850	34.00	864	0.812	20.62	288.08	428.88		
34″	850	34.00	864	0.875	22.23	309.84	461.48		
34″	850	34.00	864	0.938	23.83	331.52	493.75		
34″	850	34.00	864	1.000	25.40	352.77	525.30		
34″	850	34.00	864	1.062	26.97	373.94	556.73		
34″	850	34.00	864	1.125	28.58	395.36	588.83		
34″	850	34.00	864	1.188	30.18	416.70	620.60		
34″	850	34.00	864	1.250	31.75	437.62	651.65		
36″	900	36.00	914	0.250	6.35	95.54	142.14		
36″	900	36.00	914	0.281	7.14	107.30	159.68		
36″	900	36.00	914	0.312	7.92	119.03	176.97	10	
36″	900	36.00	914	0.344	8.74	131.12	195.12		

NPS "A" in	DN "A" mm	**O.D. in**	O.D. mm	**wall.thk "B" in**	wall.thk "B" mm	**weight lb/ft**	weight kg/m	**Sch**	**Manf. Std.**
36″	900	**36.00**	914	**0.375**	9.53	**142.81**	212.57		STD
36″	900	**36.00**	914	**0.408**	10.31	**154.48**	229.77		
36″	900	**36.00**	914	**0.438**	11.13	**166.51**	247.82		
36″	900	**36.00**	914	**0.469**	11.91	**178.14**	264.96		
36″	900	**36.00**	914	**0.500**	12.70	**189.75**	282.29	20	XS
36″	900	**36.00**	914	**0.522**	14.27	**212.90**	316.63		
36″	900	**36.00**	914	**0.625**	15.88	**236.35**	351.73	30	
36″	900	**36.00**	914	**0.688**	17.48	**259.71**	386.47		
36″	900	**36.00**	914	**0.750**	19.05	**282.62**	420.45	40	
36″	900	**36.00**	914	**0.812**	20.62	**305.44**	454.30		
36″	900	**36.00**	914	**0.875**	22.23	**328.55**	488.89		
36″	900	**36.00**	914	**0.938**	23.83	**351.57**	523.14		
36″	900	**36.00**	914	**1.000**	25.40	**374.15**	556.62		
36″	900	**36.00**	914	**1.062**	26.97	**396.64**	589.98		
36″	900	**36.00**	914	**1.125**	28.58	**419.42**	624.07		
36″	900	**36.00**	914	**1.188**	30.18	**442.10**	657.81		
36″	900	**36.00**	914	**1.250**	31.75	**464.35**	690.80		
38″	950	**38.00**	965	**0.312**	7.92	**125.70**	186.94		
38″	950	**38.00**	965	**0.344**	8.74	**138.47**	206.11		
38″	950	**38.00**	965	**0.375**	9.53	**150.83**	224.56		STD
38″	950	**38.00**	965	**0.406**	10.31	**163.16**	242.74		
38″	950	**38.00**	965	**0.438**	11.13	**175.87**	261.82		
38″	950	**38.00**	965	**0.469**	11.91	**188.17**	279.94		
38″	950	**38.00**	965	**0.500**	12.70	**200.44**	298.26		XS
38″	950	**38.00**	965	**0.562**	14.27	**224.92**	334.58		
38″	950	**38.00**	965	**0.625**	15.88	**249.71**	371.70		
38″	950	**38.00**	965	**0.688**	17.48	**274.42**	408.46		
38″	950	**38.00**	965	**0.750**	19.05	**298.65**	444.41		
38″	950	**38.00**	965	**0.812**	20.62	**322.80**	480.24		
38″	950	**38.00**	965	**0.875**	22.23	**347.26**	516.85		
38″	950	**38.00**	965	**0.938**	23.83	**371.63**	553.11		
38″	950	**38.00**	965	**1.000**	25.40	**395.53**	588.57		
38″	950	**38.00**	965	**1.062**	26.97	**419.35**	623.90		
38″	950	**38.00**	965	**1.125**	28.58	**443.47**	660.01		
38″	950	**38.00**	965	**1.188**	30.18	**467.50**	695.77		
38″	950	**38.00**	965	**1.250**	31.75	**491.07**	730.74		

NPS "A" in	DN "A" mm	O.D. in	O.D. mm	wall.thk "B" in	wall.thk "B" mm	weight lb/ft	weight kg/m	Sch	Manf. Std.
40″	1000	40.00	1016	0.312	7.92	132.37	196.90		
40″	1000	40.00	1016	0.344	8.74	145.83	217.11		
40″	1000	40.00	1016	0.375	9.53	158.85	236.54		STD
40″	1000	40.00	1016	0.406	10.31	171.84	255.71		
40″	1000	40.00	1016	0.438	11.13	185.24	275.82		
40″	1000	40.00	1016	0.469	11.91	198.19	294.92		
40″	1000	40.00	1016	0.500	12.70	211.13	314.23		XS
40″	1000	40.00	1016	0.562	14.27	236.93	352.53		
40″	1000	40.00	1016	0.625	15.88	263.07	391.67		
40″	1000	40.00	1016	0.688	17.48	289.13	430.45		
40″	1000	40.00	1016	0.750	19.05	314.69	468.37		
40″	1000	40.00	1016	0.812	20.62	340.16	506.17		
40″	1000	40.00	1016	0.875	22.23	365.97	544.81		
40″	1000	40.00	1016	0.938	23.83	391.68	583.08		
40″	1000	40.00	1016	1.000	25.40	416.91	620.51		
40″	1000	40.00	1016	1.062	26.97	442.05	657.82		
40″	1000	40.00	1016	1.125	28.58	467.52	695.96		
40″	1000	40.00	1016	1.188	30.18	492.90	722.73		
40″	1000	40.00	1016	1.250	31.75	517.80	770.67		
42″	1050	42.00	1067	0.344	8.74	153.18	228.10		
42″	1050	42.00	1067	0.375	9.53	166.86	248.53		STD
42″	1050	42.00	1067	0.406	10.31	180.52	268.67		
42″	1050	42.00	1067	0.438	11.13	194.60	289.82		
42″	1050	42.00	1067	0.469	11.91	208.22	309.90		
42″	1050	42.00	1067	0.500	12.70	221.82	330.21		XS
42″	1050	42.00	1067	0.562	14.27	248.95	370.48		
42″	1050	42.00	1067	0.625	15.88	276.44	411.64		
42″	1050	42.00	1067	0.688	17.48	303.84	452.43		
42″	1050	42.00	1067	0.750	19.05	330.72	492.33		
42″	1050	42.00	1067	0.812	20.62	357.52	532.11		
42″	1050	42.00	1067	0.875	22.23	384.67	572.77		
42″	1050	42.00	1067	0.938	23.83	411.74	613.05		
42″	1050	42.00	1067	1.000	25.40	438.29	652.46		
42″	1050	42.00	1067	1.062	26.97	464.76	691.75		
42″	1050	42.00	1067	1.125	28.58	491.57	731.91		
42″	1050	42.00	1067	1.188	30.18	518.30	771.69		
42″	1050	42.00	1067	1.250	31.75	544.52	810.80		

NPS "A" in	DN "A" mm	**O.D. in**	O.D. mm	**wall.thk "B" in**	wall.thk "B" mm	**weight lb/ft**	weight kg/m	**Sch**	**Manf. Std.**
44″	1100	**44.00**	1118	**0.344**	8.74	**160.54**	239.09		
44″	1100	**44.00**	1118	**0.375**	9.53	**174.88**	260.52		STD
44″	1100	**44.00**	1118	**0.406**	10.31	**189.20**	281.64		
44″	1100	**44.00**	1118	**0.438**	11.13	**203.97**	303.82		
44″	1100	**44.00**	1118	**0.489**	11.91	**218.25**	324.88		
44″	1100	**44.00**	1118	**0.500**	12.70	**232.51**	346.18		XS
44″	1100	**44.00**	1118	**0.562**	14.27	**260.97**	388.42		
44″	1100	**44.00**	1118	**0.625**	15.88	**289.80**	431.62		
44″	1100	**44.00**	1118	**0.688**	17.48	**318.55**	474.42		
44″	1100	**44.00**	1118	**0.750**	19.05	**346.76**	512.29		
44″	1100	**44.00**	1118	**0.812**	20.62	**374.88**	558.04		
44″	1100	**44.00**	1118	**0.875**	22.23	**403.38**	600.73		
44″	1100	**44.00**	1118	**0.938**	23.83	**431.79**	643.03		
44″	1100	**44.00**	1118	**1.000**	25.40	**458.67**	684.41		
44″	1100	**44.00**	1118	**1.062**	26.97	**487.47**	725.67		
44″	1100	**44.00**	1118	**1.125**	28.58	**515.63**	767.85		
44″	1100	**44.00**	1118	**1.188**	30.18	**543.70**	809.95		
44″	1100	**44.00**	1118	**1.250**	31.75	**571.25**	850.54		
46″	1150	**46.00**	1168	**0.344**	8.74	**167.89**	249.87		
46″	1150	**46.00**	1168	**0.375**	9.53	**182.90**	272.27		STD
46″	1150	**46.00**	1168	**0.406**	10.31	**197.98**	294.35		
46″	1150	**46.00**	1168	**0.438**	11.13	**213.33**	317.54		
46″	1150	**46.00**	1168	**0.469**	11.91	**228.27**	339.56		
46″	1150	**46.00**	1168	**0.500**	12.70	**243.20**	361.84		XS
46″	1150	**46.00**	1168	**0.562**	14.27	**272.98**	406.02		
46″	1150	**46.00**	1168	**0.625**	15.98	**303.16**	451.20		
46″	1150	**46.00**	1168	**0.688**	17.48	**333.26**	495.97		
46″	1150	**46.00**	1168	**0.750**	19.05	**362.79**	539.78		
46″	1150	**46.00**	1168	**0.812**	20.62	**392.24**	583.47		
46″	1150	**46.00**	1168	**0.875**	22.23	**422.09**	628.14		
46″	1150	**46.00**	1168	**0.938**	23.83	**451.85**	672.41		
46″	1150	**46.00**	1168	**1.000**	25.40	**481.05**	716.73		
46″	1150	**46.00**	1168	**1.062**	26.97	**510.17**	758.92		
46″	1150	**46.00**	1168	**1.125**	28.58	**539.68**	803.09		
46″	1150	**46.00**	1168	**1.188**	30.18	**569.10**	846.86		
46″	1150	**46.00**	1168	**1.250**	31.75	**597.97**	889.69		
48″	1200	**48.00**	1219	**0.344**	8.74	**175.25**	260.86		
48″	1200	**48.00**	1219	**0.375**	9.53	**190.02**	284.25		STD
48″	1200	**48.00**	1219	**0.406**	10.31	**206.56**	307.32		
48″	1200	**48.00**	1219	**0.438**	11.13	**222.70**	331.54		

NPS "A" in	DN "A" mm	O.D. in	O.D. mm	wall.thk "B" in	wall.thk "B" mm	weight lb/ft	weight kg/m	Sch	Manf. Std.
48″	1200	**48.00**	1219	**0.469**	11.91	**238.30**	354.54		
48″	1200	**48.00**	1219	**0.500**	12.70	**253.89**	377.81		XS
48″	1200	**48.00**	1219	**0.562**	14.27	**285.00**	423.97		
48″	1200	**48.00**	1219	**0.625**	15.88	**316.52**	471.17		
48″	1200	**48.00**	1219	**0.688**	17.48	**347.97**	517.95		
48″	1200	**48.00**	1219	**0.750**	19.05	**378.83**	563.74		
48″	1200	**48.00**	1219	**0.812**	20.62	**409.61**	609.40		
48″	1200	**48.00**	1219	**0.875**	22.23	**440.80**	656.10		
48″	1200	**48.00**	1219	**0.938**	23.83	**471.90**	702.38		
48″	1200	**48.00**	1219	**1.000**	25.40	**502.43**	747.67		
48″	1200	**48.00**	1219	**1.062**	26.97	**532.88**	792.84		
48″	1200	**48.00**	1219	**1.125**	28.58	**563.73**	839.04		
48″	1200	**48.00**	1219	**1.188**	30.18	**594.50**	884.82		
48″	1200	**48.00**	1219	**1.250**	31.75	**624.70**	929.62		
52″	1300	**52.00**	1321	**0.375**	9.53	**206.95**	308.23		
52″	1300	**52.00**	1321	**0.406**	10.31	**223.93**	333.26		
52″	1300	**52.00**	1321	**0.438**	11.13	**241.42**	359.54		
52″	1300	**52.00**	1321	**0.469**	11.91	**258.36**	384.50		
52″	1300	**52.00**	1321	**0.500**	12.70	**275.27**	409.76		
52″	1300	**52.00**	1321	**0.562**	14.27	**309.03**	459.86		
52″	1300	**52.00**	1321	**0.625**	15.88	**343.25**	511.12		
52″	1300	**52.00**	1321	**0.688**	17.48	**377.39**	561.93		
52″	1300	**52.00**	1321	**0.750**	19.05	**410.90**	611.66		
52″	1300	**52.00**	1321	**0.812**	20.62	**444.33**	661.27		
52″	1300	**52.00**	1321	**0.875**	22.23	**478.21**	712.02		
52″	1300	**52.00**	1321	**0.938**	23.83	**512.01**	762.33		
52″	1300	**52.00**	1321	**1.000**	25.40	**545.19**	811.57		
52″	1300	**52.00**	1321	**1.062**	26.97	**578.29**	860.69		
52″	1300	**52.00**	1321	**1.125**	28.58	**611.84**	910.93		
52″	1300	**52.00**	1321	**1.188**	30.18	**645.30**	960.74		
52″	1300	**52.00**	1321	**1.250**	31.75	**678.15**	1009.49		
56″	1400	**56.00**	1422	**0.375**	9.53	**222.99**	331.96		
56″	1400	**56.00**	1422	**0.406**	10.31	**241.29**	358.94		
56″	1400	**56.00**	1422	**0.438**	11.13	**260.15**	387.26		
56″	1400	**56.00**	1422	**0.469**	11.91	**278.41**	414.17		
56″	1400	**56.00**	1422	**0.500**	12.70	**296.65**	441.39		
56″	1400	**56.00**	1422	**0.562**	14.27	**333.06**	495.41		
56″	1400	**56.00**	1422	**0.625**	15.88	**369.97**	550.67		
56″	1400	**56.00**	1422	**0.688**	17.48	**406.80**	605.46		

NPS "A" in	DN "A" mm	**O.D. in**	O.D. mm	**wall.thk "B" in**	wall.thk "B" mm	**weight lb/ft**	weight kg/m	**Sch**	**Manf. Std.**
56″	1400	**56.00**	1422	**0.750**	19.05	**442.97**	659.11		
56″	1400	**56.00**	1422	**0.812**	20.62	**479.05**	712.63		
56″	1400	**56.00**	1422	**0.875**	22.23	**515.63**	767.39		
56″	1400	**56.00**	1422	**0.938**	23.83	**552.12**	821.68		
56″	1400	**56.00**	1422	**1.000**	25.40	**587.95**	874.83		
56″	1400	**56.00**	1422	**1.062**	26.97	**623.70**	927.86		
56″	1400	**56.00**	1422	**1.125**	28.58	**659.94**	982.12		
56″	1400	**56.00**	1422	**1.188**	30.18	**696.10**	1035.91		
56″	1400	**56.00**	1422	**1.250**	31.75	**731.60**	1088.57		
60″	1500	**60.00**	1524	**0.375**	9.53	**239.02**	355.94		
60″	1500	**60.00**	1524	**0.406**	10.31	**258.65**	384.87		
60″	1500	**60.00**	1524	**0.438**	11.13	**278.88**	415.26		
60″	1500	**60.00**	1524	**0.469**	11.91	**296.47**	444.13		
60″	1500	**60.00**	1524	**0.500**	12.70	**318.03**	473.34		
60″	1500	**60.00**	1524	**0.562**	14.27	**357.09**	531.30		
60″	1500	**60.00**	1524	**0.625**	15.88	**396.70**	590.62		
60″	1500	**60.00**	1524	**0.688**	17.48	**436.22**	649.44		
60″	1500	**60.00**	1524	**0.750**	19.05	**475.04**	707.03		
60″	1500	**60.00**	1524	**0.812**	20.62	**513.77**	764.50		
60″	1500	**60.00**	1524	**0.875**	22.23	**553.04**	823.31		
60″	1500	**60.00**	1524	**0.938**	23.83	**592.23**	881.63		
60″	1500	**60.00**	1524	**1.000**	25.40	**630.71**	938.73		
60″	1500	**60.00**	1524	**1.062**	26.97	**669.11**	995.71		
60″	1500	**60.00**	1524	**1.125**	28.58	**708.05**	1054.01		
60″	1500	**60.00**	1524	**1.188**	30.18	**746.90**	1111.83		
60″	1500	**60.00**	1524	**1.250**	31.75	**785.05**	1168.44		
64″	1600	**64.00**	1626	**0.375**	9.53	**255.06**	379.91		
64″	1600	**64.00**	1626	**0.406**	10.31	**276.01**	410.81		
64″	1600	**64.00**	1626	**0.438**	11.13	**297.61**	443.25		
64″	1600	**64.00**	1626	**0.469**	11.91	**318.52**	474.09		
64″	1600	**64.00**	1626	**0.500**	12.70	**338.41**	505.29		
64″	1600	**64.00**	1626	**0.562**	14.27	**381.12**	567.20		
64″	1600	**64.00**	1626	**0.625**	15.88	**423.42**	630.56		
64″	1600	**64.00**	1626	**0.688**	17.48	**485.64**	693.41		
64″	1600	**64.00**	1626	**0.750**	19.05	**507.11**	754.95		
64″	1600	**64.00**	1626	**0.812**	20.62	**548.49**	816.37		
64″	1600	**64.00**	1626	**0.875**	22.23	**590.46**	879.23		
64″	1600	**64.00**	1626	**0.938**	23.83	**632.34**	941.57		

NPS "A" in	DN "A" mm	O.D. in	O.D. mm	wall.thk "B" in	wall.thk "B" mm	weight lb/ft	weight kg/m	Sch	Manf. Std.
64″	1600	**64.00**	1626	**1.000**	25.40	**673.47**	1002.62		
64″	1600	**64.00**	1626	**1.062**	26.97	**714.52**	1063.55		
64″	1600	**64.00**	1626	**1.125**	28.58	**756.15**	1125.90		
64″	1600	**64.00**	1626	**1.188**	30.18	**797.69**	1187.74		
64″	1600	**64.00**	1626	**1.250**	31.75	**838.50**	1248.30		
68″	1700	**68.00**	1727	**0.469**	11.91	**338.57**	503.75		
68″	1700	**68.00**	1727	**0.500**	12.70	**360.79**	536.92		
68″	1700	**68.00**	1727	**0.562**	14.27	**405.15**	602.71		
68″	1700	**68.00**	1727	**0.625**	15.88	**450.15**	670.12		
68″	1700	**68.00**	1727	**0.688**	17.48	**495.06**	736.95		
68″	1700	**68.00**	1727	**0.750**	19.05	**539.18**	802.40		
68″	1700	**68.00**	1727	**0.812**	20.62	**583.21**	867.73		
68″	1700	**68.00**	1727	**0.875**	22.23	**627.87**	934.60		
68″	1700	**68.00**	1727	**0.938**	23.83	**672.45**	1000.92		
68″	1700	**68.00**	1727	**1.000**	25.40	**716.23**	1065.89		
68″	1700	**68.00**	1727	**1.062**	26.97	**759.93**	1130.73		
68″	1700	**68.00**	1727	**1.125**	28.58	**804.26**	1197.09		
68″	1700	**68.00**	1727	**1.188**	30.16	**848.49**	1262.92		
68″	1700	**68.00**	1727	**1.250**	31.75	**891.95**	1327.39		
72″	1800	**72.00**	1829	**0.500**	12.70	**382.17**	568.87		
72″	1800	**72.00**	1829	**0.562**	14.27	**429.18**	638.64		
72″	1800	**72.00**	1829	**0.625**	15.88	**476.87**	710.06		
72″	1800	**72.00**	1829	**0.688**	17.48	**524.48**	780.92		
72″	1800	**72.00**	1829	**0.750**	19.05	**571.25**	850.32		
72″	1800	**72.00**	1829	**0.812**	20.62	**617.93**	919.60		
72″	1800	**72.00**	1829	**0.875**	22.23	**665.29**	990.52		
72″	1800	**72.00**	1829	**0.938**	23.83	**712.55**	1060.87		
72″	1800	**72.00**	1829	**1.000**	25.40	**758.99**	1129.78		
72″	1800	**72.00**	1829	**1.062**	26.97	**805.34**	1198.57		
72″	1800	**72.00**	1829	**1.125**	28.58	**852.36**	1268.98		
72″	1800	**72.00**	1829	**1.188**	30.18	**899.29**	1338.83		
72″	1800	**72.00**	1829	**1.250**	31.75	**945.40**	1407.25		
76″	1900	**76.00**	1930	**0.500**	12.70	**403.55**	600.50		
76″	1900	**76.00**	1930	**0.562**	14.27	**453.21**	674.18		
76″	1900	**76.00**	1930	**0.625**	15.88	**503.60**	749.62		
76″	1900	**76.00**	1930	**0.688**	17.48	**553.90**	824.45		
76″	1900	**76.00**	1930	**0.750**	19.05	**603.32**	897.77		
76″	1900	**76.00**	1930	**0.812**	20.62	**652.65**	970.96		
76″	1900	**76.00**	1930	**0.875**	22.23	**702.70**	1045.89		
76″	1900	**76.00**	1930	**0.938**	23.83	**752.66**	1120.22		

NPS "A" in	DN "A" mm	O.D. in	O.D. mm	wall.thk "B" in	wall.thk "B" mm	weight lb/ft	weight kg/m	Sch	Manf. Std.
76″	1900	**76.00**	1930	**1.000**	25.40	**801.75**	1193.05		
76″	1900	**76.00**	1930	**1.062**	26.97	**850.75**	1265.74		
76″	1900	**76.00**	1930	**1.125**	28.58	**900.47**	1340.17		
76″	1900	**76.00**	1930	**1.188**	30.18	**950.09**	1414.01		
76″	1900	**76.00**	1930	**1.250**	31.75	**998.85**	1486.33		
80″	2000	**80.00**	2032	**0.562**	14.27	**477.25**	710.08		
80″	2000	**80.00**	2032	**0.625**	15.88	**530.32**	789.56		
80″	2000	**80.00**	2032	**0.688**	17.48	**583.32**	868.43		
80″	2000	**80.00**	2032	**0.750**	19.05	**635.39**	945.69		
80″	2000	**80.00**	2032	**0.812**	20.62	**687.37**	1022.83		
80″	2000	**80.00**	2032	**0.875**	22.23	**740.12**	1101.81		
80″	2000	**80.00**	2032	**0.938**	23.83	**792.77**	1180.17		
80″	2000	**80.00**	2032	**1.000**	25.40	**844.51**	1256.94		
80″	2000	**80.00**	2032	**1.062**	26.97	**896.17**	1333.59		
80″	2000	**80.00**	2032	**1.125**	28.58	**948.57**	1412.06		
80″	2000	**80.00**	2032	**1.188**	30.18	**1000.89**	1489.92		
80″	2000	**80.00**	2032	**1.250**	31.75	**1052.30**	1566.20		

7. STAINLESS STEEL PIPE (SEAMLESS AND WELDED), DIMENSIONS AND WEIGHT

Dimensions—as per ASME B36.19.

Weights—manufacturer's estimates.

Size range—$^{1}/_{8}$–30 in. outside diameter (OD).

Wall thickness (WT)—5S, 10S, 40S, and 80S.

Methods of manufacture—covered in the relevant ASTM specification.

Chemical composition and mechanical properties—covered in the relevant ASTM Specification.

Tolerances and permissible variations—depend on the method of manufacture, which are covered in the relevant ASTM specification.

Pipe length—covered in the relevant ASTM specification.

NPS A (in.)	DN A (mm)	OD (in.)	OD (mm)	WT B (in.)	WT B (mm)	Weight (lb/ft)	Weight (kg/m)	Schedule
1/8		0.405	10.3	0.049	1.24	0.19	0.28	10S
1/8		0.405	10.3	0.068	1.73	0.24	0.37	40S
1/8		0.405	10.3	0.095	2.41	0.31	0.47	80S
1/4		0.540	13.7	0.065	1.65	0.33	0.49	10S
1/4		0.540	13.7	0.088	2.24	0.42	0.63	40S
1/4		0.540	13.7	0.119	3.02	0.54	0.80	80S
3/8	10	0.675	17.1	0.065	1.65	0.42	0.63	10S
3/8	10	0.675	17.1	0.091	2.31	0.57	0.84	40S
3/8	10	0.675	17.1	0.126	3.20	0.74	1.10	80S
1/2	15	0.840	21.3	0.065	1.65	0.54	0.80	5S
1/2	15	0.840	21.3	0.083	2.11	0.67	1.00	10S
1/2	15	0.840	21.3	0.109	2.77	0.85	1.27	40S
1/2	15	0.840	21.3	0.146	3.73	1.09	1.62	80S
3/4	20	1.050	26.7	0.065	1.65	0.69	1.03	5S
3/4	20	1.050	26.7	0.083	2.11	0.86	1.28	10S
3/4	20	1.050	26.7	0.113	2.87	1.13	1.69	40S
3/4	20	1.050	26.7	0.154	3.91	1.47	2.20	80S
1	25	1.315	33.4	0.065	1.65	0.87	1.30	5S
1	25	1.315	33.4	0.109	2.77	1.40	2.09	10S
1	25	1.315	33.4	0.133	3.38	1.68	2.50	40S
1	25	1.315	33.4	0.179	4.55	2.17	3.24	80S
1 1/4	32	1.660	42.2	0.065	1.65	1.11	1.65	5S
1 1/4	32	1.660	42.2	0.109	2.77	1.81	2.70	10S
1 1/4	32	1.660	42.2	0.140	3.56	2.27	3.39	40S
1 1/4	32	1.660	42.2	0.191	4.85	3.00	4.47	80S
1 1/2	40	1.900	48.3	0.065	1.65	1.28	1.91	5S
1 1/2	40	1.900	48.3	0.109	2.77	2.09	3.11	10S
1 1/2	40	1.900	48.3	0.145	3.68	2.72	4.05	40S
1 1/2	40	1.900	48.3	0.200	5.08	3.63	5.41	80S
2	50	2.375	60.3	0.065	1.65	1.61	2.40	5S
2	50	2.375	60.3	0.109	2.77	2.64	3.93	10S
2	50	2.375	60.3	0.154	3.91	3.65	5.44	40S
2	50	2.375	60.3	0.218	5.54	5.02	7.48	80S

NPS A (in.)	DN A (mm)	OD (in.)	OD (mm)	WT B (in.)	WT B (mm)	Weight (lb/ft)	Weight (kg/m)	Schedule
2½	65	2.875	73.0	0.083	2.11	2.48	3.69	5S
2½	65	2.875	73.0	0.120	3.05	3.53	5.26	10S
2½	65	2.875	73.0	0.203	5.16	5.79	8.63	40S
2½	65	2.875	73.0	0.276	7.01	7.66	11.41	80S
3	75	3.500	0.083	0.083	2.11	3.03	4.51	5S
3	75	3.500	0.083	0.120	3.05	4.33	6.45	10S
3	75	3.500	0.083	0.216	5.49	7.58	11.29	40S
3	75	3.500	0.083	0.300	7.62	10.25	15.27	80S
3½		4.000	101.6	0.083	2.11	3.48	5.18	5S
3½		4.000	101.6	0.120	3.05	4.97	7.40	10S
3½		4.000	101.6	0.226	5.74	9.11	13.57	40S
3½		4.000	101.6	0.318	8.08	12.50	18.63	80S
4	100	4.500	114.3	0.083	2.11	3.92	5.84	5S
4	100	4.500	114.3	0.120	3.05	5.61	8.36	10S
4	100	4.500	114.3	0.237	6.02	10.79	16.07	40S
4	100	4.500	114.3	0.337	8.56	14.98	22.32	80S
5	125	5.563	141.3	0.109	2.77	6.36	9.47	5S
5	125	5.563	141.3	0.134	3.40	7.77	11.57	10S
5	125	5.563	141.3	0.258	6.55	14.62	21.77	40S
5	125	5.563	141.3	0.375	9.53	20.78	30.97	80S
6	150	6.625	168.3	0.109	2.77	7.60	11.32	5S
6	150	6.625	168.3	0.134	3.40	9.29	13.84	10S
6	150	6.625	168.3	0.280	7.11	18.97	28.26	40S
6	150	6.625	168.3	0.432	10.97	28.57	42.56	80S
8	200	8.625	219.1	0.109	2.77	9.93	14.79	5S
8	200	8.625	219.1	0.148	3.76	13.40	19.96	10S
8	200	8.625	219.1	0.322	8.18	28.55	42.55	40S
8	200	8.625	219.1	0.500	12.70	43.39	64.64	80S
10	250	10.750	273.0	0.134	3.40	15.19	22.63	5S
10	250	10.750	273.0	0.165	4.19	18.65	27.78	10S
10	250	10.750	273.0	0.365	9.27	40.48	60.31	40S
10	250	10.750	273.0	0.500	12.70	54.74	81.55	80S

Note: Schedules 5S and 10S wall thickness do not permit threading in accordance with ASME B1.20.1.

(Continues)

(Continued)

NPS A (in.)	DN A (mm)	OD (in.)	OD (mm)	WT B (in.)	WT B (mm)	Weight (lb/ft)	Weight (kg/m)	Schedule
12	300	12.750	323.8	0.156	3.95	20.98	31.25	5S
12	300	12.750	323.8	0.180	4.57	24.17	36.00	10S
12	300	12.750	323.8	0.375	9.53	49.56	73.88	40S
12	300	12.750	323.8	0.500	12.70	65.42	97.46	80S
14	350	14.000	355.6	0.156	3.96	23.07	34.36	5S
14	350	14.000	355.6	0.188	4.78	27.73	41.35	10S
16	400	16.00	406.4	0.165	4.19	27.90	41.56	5S
16	400	16.00	406.4	0.188	4.78	31.75	47.34	10S
18	450	18.00	457.0	0.165	4.19	31.43	46.81	5S
18	450	18.00	457.0	0.188	4.78	35.76	53.31	10S
20	500	20.00	508	0.188	4.78	39.78	59.25	5S
20	500	20.00	508	0.218	5.54	46.27	68.89	10S
22	550	22.00	559	0.188	4.78	43.80	65.24	5S
22	550	22.00	559	0.218	5.54	50.71	75.53	10S
24	600	24.00	610	0.218	5.54	55.37	82.47	5S
24	600	24.00	610	0.250	6.35	63.41	94.53	10S
30	750	30.00	762	0.250	6.35	79.43	118.31	5S
30	750	30.00	762	0.312	7.92	98.93	132.91	10S

Note: Schedules 5S and 10S wall thickness do not permit threading in accordance with ASME B1.20.1.

5

JOINTS FOR PROCESS PIPING SYSTEMS

1. INTRODUCTION TO PIPE JOINTS

A piping systems forms the arteries through which a process fluid flows, and this pipe connects the various pieces of equipment that are required, within a plant, to refine the product. To facilitate changes of direction and regulate the flow, these straight lengths of pipe must be connected to piping components, valves and process equipment, to complete the system. Numerous options are available to the piping engineer responsible for specifying the correct method of jointing for a particular process piping system.

Listed here are several alternatives for joints to be used within metal piping systems:

- Flanged—using weld neck, socket weld, screwed, lap joint flanges.
- Butt weld—using a full-penetration weld.
- Socket weld—using a fillet weld with socket weld couplings.
- Screwed—using screwed couplings.
- Hubbed connections—using propriety hubs and collars.
- Mechanical coupling—victaullic-type couplings.
- Soldered.

All these methods have potential leak paths, and careful consideration must be made during the material selection process and in the choice of level of inspection to minimize "in-service" fluid loss.

The joint type chosen must be leak free for the duration of the plant life. Therefore, the following factors must be taken into consideration:

- Type of process fluid—its toxicity and viscosity.
- Design temperature range.
- Design pressure.
- Mechanical strength of the base material—its tensile strength and ability to yield.
- Size.
- Weight.
- Erosion and corrosion
- For permanent or temporary use, need for quick release.
- Quality of the labor available.
- Cost.
- Maintainability and reliability.
- Plant life.
- Need to handle vibration.
- External mechanical impact from personnel, vehicles, and the like.
- Ease of fabrication or erection.
- Availability.

This is a large checklist; however, many materials immediately are prohibited after the first four points are considered.

Many piping systems have more than one type of pipe jointing to suit the plant's requirements. However, piping systems always are limited by the method of jointing considered to be the least efficient. Generally, this is the mating of two flanges with a set of bolts and a gasket compressed under calculated bolt loads.

The purpose of this chapter is to assist you in evaluating the type of joint most suitable after having evaluated all these factors.

2. FUNDAMENTAL PRINCIPLES

The piping joint selected must maintain the integrity of the complete piping system of which it is a part. The joint must not leak while it is in service, and it may be subjected to both internal and external loadings.

We next examine some factors to consider when choosing a type of pipe joint.

Type of Process Fluid

The type of fluid to be transported must first be considered:

- Hazardous process—see ASME 31.3, Category M.
- Nonhazardous process—see ASME B31.3, normal fluid service (NFS).
- Utility service—see ASME 31.3, Category D.

Some process fluids, such as ammonia and concentrated acids, are defined as hazardous, and even the smallest leakage is considered dangerous to personnel and the plant. In these cases, the piping joint chosen is the one that is most efficient, regardless of cost. This joint is a butt weld, which offers the best option and the one least likely to fail, especially when supported by a strict inspection regime.

Pressure and Temperature

Flanged joints are considered the joint with the lowest integrity, and they are used as the basis to set the upper design limit of a piping system.

Tabulated data in ASME B16.5 for steel flanges states the maximum allowable internal design pressure for a specific material in a piping class at a given temperature. This allowable internal pressure reduces as the temperature increases.

ASME B16.5 covers pipe flanges from 1/2 to 24 in., and for flanges 26 in. and above, reference is made to ASME B16.47, series A and B. Both of these are dimensional standards, and time should be taken to review them thoroughly to understand the full scope of these two documents.

The flange class for both standards are as follows:

- 150 lb.
- 300 lb.
- 400 lb.
- 600 lb.
- 900 lb.

- 1500 lb.
- 2500 lb.

The designation is the maximum pressure that the flange is "rated" to at an elevated temperature. Another term for class is *rating*.

Example 1. See Material Group Index, Group 1, extracted from ASME B16.5:

Material: ASTM A105N–ASME B16.5, Material Group 1.1.
Piping class: ASME B16.5, Class 150.
Design temperature: 100°F.
Allowable internal pressure: 285 psi This figure limits the use of these flanges to a design pressure of psi, the working or operating pressure is marginally lower than this figure. The piping system will be subjected to a hydrostatic test pressure of 1.5 times the design pressure, to test the integrity of the fabrication welds.

Example 2. See Material Group Index, Group 1, extracted from ASME B16.5:

Material: ASTM A105N–ASME B16.5-Material Group 1.1.
Piping Class: ASME B16.5, Class 150.
Design Temperature: 300°F.
Allowable internal pressure: 230 psi

Butt weld joints are considered to have the highest integrity; and a full penetration butt weld that has been inspected using either radiography (RT) or ultrasonics (UT) is considered to be guaranteed leakproof. Other methods of nondestructive examination, such as magnetic particle examination (MPE) or liquid penetration examination (LPE), for nonmagnetic metals, come a very close second. Piping systems carrying toxic fluids or operating under very high pressures and temperatures may be subjected to 100% NDE, which means that all valves are X-rayed.

This NDE takes place before the hydrostatic testing of a piping system. Once a hyrotest has been carried out successfully to 1.5 times the design pressure, all welds are considered to be of the highest integrity.

Socket weld connections are fillet welds, which although not full penetration welds, are considered by most operators suitable for handling process fluid. For added confidence, they can also be subjected to NDE, such as RT, MPE, or LPE for nonmagnetic metals.

Care must be taken with the fit up of socket weld connections. A gap must be left at the bottom of the female socket to prevent "bottoming" during the welding process, when heat is applied and the metal expands.

Screwed connections are not suitable for conditions in which fluid experiences both high temperature and high pressures or is subjected to vibration. A screwed connection, however, is capable of containing medium to high pressure, but because of its lower integrity, many operators restrict the use to utility piping systems, such as air, water, and nitrogen. Also, piping systems transporting toxic fluids require high-integrity pipe joints such as butt welds. I cover the strength or weaknesses of specific pipe joints at a later stage.

Material Compatibility. The material used for the pipe joint must be mechanically and chemically compatible with the pipe transporting the fluid. If welding is required, then the two materials must also be chemically compatible to effect a correct weld. Further, the material of construction of the joint must have very close corrosion-resistant characteristics to the parent pipe, for the fluid transported internally and the external environment. For use in food and drug industries, the jointing material must not contaminate the process fluid.

Materials of differing chemical compositions can be welded together as long as there is no possibility of galvanic corrosion, the correct weld procedure is in place, and the weld is executed by a suitably qualified technician.

Size. Some joints are limited by the outside diameter of the pipe. Screwed fittings can be used in diameters up to 4 in. (100 DN), but in practice, they are rarely used above 2 in. (50 DN). Socket weld fittings, when specified, are usually used only in diameters up to 2 in. (50 DN). Butt welded and flanged joints can be used from ½ in. (13 DN) to as high as is feasibly possible.

Weight and Space. The weight and the space taken up by the joint may need to be considered. Flanged joints in the higher ASME piping classes take up a great deal of space and weigh a considerable amount. For offshore projects, this may have to be considered, if space is at a premium and the weight of the piping system must be taken into consideration for module lifts, when they are installed at sea.

Corrosion. When coupled, screwed pipe joints create very small crevices, and this is not advisable with certain process fluids at extreme pressure or temperature conditions. Over an extended period, such crevices can accelerate corrosion, which reduces the efficiency of the joint and may lead to in-service failure. External corrosion from the environment—hot, as in the desert; cold, as in Alaska; or wet, as in marine conditions—must also be considered.

Permanent or Temporary Connection. If the connection joins piece of pipe to a valve or a piece of equipment and this joint might have to be broken to remove the item, then a flanged joint or a mechanical coupling should be considered. A welded joint is not an option, because it is considered permanent. Some valves are designed for top entry, which means that they can be repaired and maintained in situ and therefore may be fully welded into the line, because they need not be removed.

When pipe connections are broken on a regular basis, as in the case of flexible hose connections to hard pipe, then a quick-release alternative should be considered.

Loadings. A joint must be leakproof when all operating and external loads have been considered. These forces—axial (tensile and compressive), shear, torsion, and bending—occur when the plant is operating with fluctuating temperatures and pressures. Stress analysis and adequate pipe supporting help to distribute these forces. However, certain joints, such as screwed connections, are unsuitable for extreme pressure and temperature conditions; and they must be excluded.

Quality of Fabrication and Erection Labor. The workforce employed to fabricate and erect the piping system must be able to competently execute the work in accordance with the relevant international codes to which the plant has been designed. This particularly applies to welding, and suitable qualified personnel must be tested to guarantee that they can perform the welds required with the minimum level of failure during the testing and commissioning of the plant.

Cost. All joints have different costs based on the mating components and the labor required to complete the connection.

The following examples look at a variety of joints connecting two straight lengths of pipe. Each requires different components and a specific procedure carried out by labor trained and qualified to completed the joint successfully. Of these, the first method is the cheapest and the final one the most expensive.

Pipe to Pipe, Screwed

Material: One screwed joint coupling female ends.
Labor: No welds, the joint executed by a fitter.

Pipe to Pipe, Socket Weld

Material: One socket weld coupling female ends.
Labor: Two fillet welds executed by a welder.

Pipe to Pipe, Butt Weld

Material: Two pipes with prepared ends.
Labor: One butt weld executed by a welder.

Pipe to Pipe, Flanged (threaded)

Material: Two threaded flanges, one gasket, one set of stud bolts.
Labor: Two threaded connections and one flanged connection executed by a fitter.

Pipe to Pipe, Flanged (socket weld)

Material: Two socket weld flanges, one gasket, one set of stud bolts.
Labor: Two fillet welded connections and one flanged connection executed by a welder and a fitter.

Pipe to Pipe, Flanged (butt weld)

Material: Two weld neck flanges, one gasket, one set of stud bolts.

Labor: Two butt weld connections and one flanged connection executed by a fitter.

Maintainability. Butt welds, socket welds, and screwed connections do not require maintenance unless there is a failure in the joint. However, temporary flanged joints require a new gaskets every time the joint is broken; and if this occurs regularly, the bolting has to be changed out.

Vibration. Certain items of mechanical equipment are subject to vibration, especially pumps and compressors. The jointing of piping systems hooked up to this equipment or run in close proximity must be able to withstand this constant movement without failing. Screwed connections are often prohibited on piping systems located close to such items of equipment.

Conclusion. All of these points must be considered when selecting a suitable pipe joint. In many cases, certain methods of jointing are immediately rejected, and it therefore becomes unnecessary to consider the other options. The plant owner or client sometimes prohibits the use of certain joints, based on the experience with in-service failure in other plants. This helps the piping engineer during the selection.

Many clients also have corporate specifications for process and utility piping systems, and these documents can be used as technical templates to create project-specific documents.

The final selection can be made only after careful consideration of the following factors: design conditions, construction, commissioning, operation conditions, and plant life.

3. WELDED JOINTS

Welding is a relatively cheap method of joining two metallic components—pipe to pipe, pipe to fitting, or fitting to fitting—to create an effective pressure seal. This joint can be inspected using nondestructive examination (NDE) and hydrostatically tested to satisfy the relevant codes.

Welding Carbon Steel and Low-Alloy Carbon Steel

Carbon steel and low alloy carbon steel can be divided into five major groups: carbon steel, high-strength low-alloy steels, quenched and tempered steels, heat-treatable low-alloy steels, and chromium-molybdenum steels.

Steel is considered a carbon steel when no minimum content is specified for chromium, cobalt, columbium (niobium), molybdenum, nickel, titanium, tungsten, vanadium, zirconium, or any other alloying element.

Welding Stainless Steel

The stainless properties are due to the presence of chromium in quantities greater than 12% by weight. This level of chromium is the minimum level to ensure a continuous stable layer.

Stainless steel can be divided into three major groups: austenitic (300 series), ferritic, martensitic (400 series), and ferritic-austenitic (duplex). Two types of metallic welding (butt welding and socket welding) can be used to join straight lengths of steel pipe, pipe to fitting, or fitting to fitting. Both methods have their advantages and disadvantages, which are highlighted in the following table:

Type of Weld	Characteristics	Disadvantages
Butt weld (full penetration)	High integrity Suitable for elevated-temperature and high-pressure service Small and large sizes All wall thicknesses	Expensive Requires skilled labor Time consuming
Socket weld (fillet weld)	Medium integrity Suitable for high temperature and medium pressure 2 in. and smaller Cheaper to fabricate	Limited to sizes 2 in. and smaller Not suitable for high pressure

Butt Welds

A joint is butt welded when two pieces of pipe or fittings are supplied with matching beveled ends, butted together, held firmly in position, then welded, using a specific welding procedure and completed by a suitably qualified person.

The welding procedure specification (WPS) covers the following:

- Pipe material, diameter, and wall thickness.
- Joint preparation.
- Position of pipe (vertical or horizontal).
- Back purging gas (if applicable).
- Preheating and interpass temperatures.
- Type of welding process.
- Flux and shielding gas.
- Electrode and filler material.
- Gas flow rate and nozzle details.
- Welding current (ac, dc, polarity).
- Postweld heat treatment.
- Identification of the welder.

There are three types of butt welds: Full penetration, with a backing ring, and with a fusible backing ring. Of these, by far the most common in the oil and gas industry is the first, the full-penetration butt weld without a backing ring. If executed by suitably qualified personnel, using the correct WPS, it results in a high-integrity, pressure-retaining weld that can be subjected to NDE for added confidence.

Socket Welds

To join two square-cut pieces of pipe, a socket weld coupling is required (see Figure 5.1). A socket-weld coupling allows the two pipe lengths to be inserted into the ends of the fitting and the two circumferential fillet welds completed. A root gap of approximately 1.5 mm is required to accommodate lateral expansion of the pipe when heat is applied during the welding process (see Figure 5.2). If this gap is omitted, as the pipe expands, "bottoming" takes place at the base of the socket, which results in unnecessary force applied to the joint during the welding process.

Socket-weld joints are economical up to about 2 in. (50 mm); however, at sizes larger than this, the higher-integrity butt weld becomes commercially viable. Joining two pieces of pipe using the socket-weld method requires two fillet welds and a full coupling. The butt-weld method requires only one full-penetration weld and no additional fitting and results in a higher-integrity weld.

4. WELDING TECHNIQUES

For the purposes or this book, we cover weld process generally employed in the fabrication and erection of piping systems for the oil and gas industry.

There are two methods of applying a weld: The manual method generally is used for shop and site work; the semi-automatic and

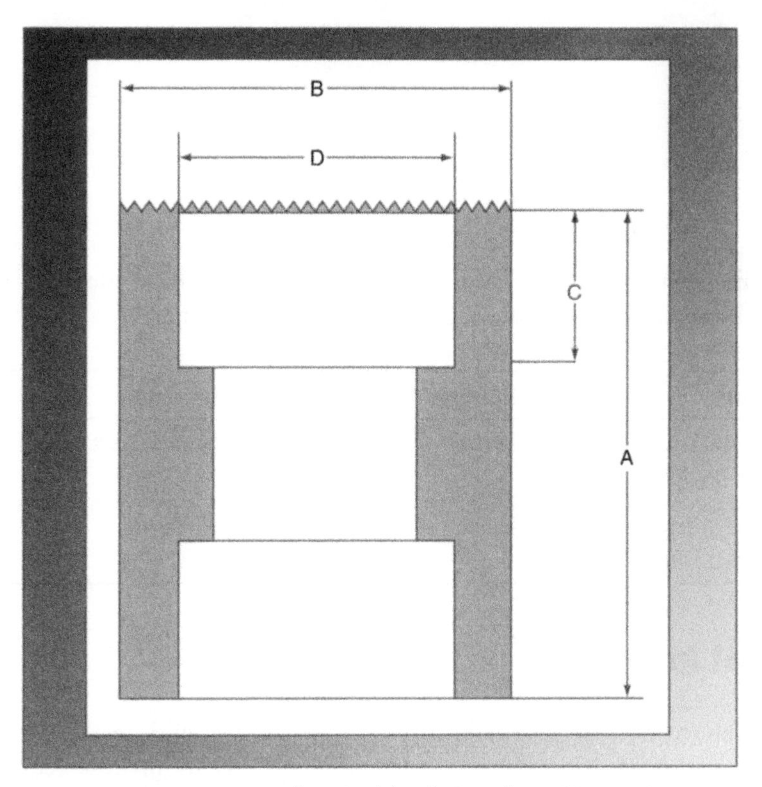

Figure 5.1. Socket Weld Full Coupling. (Section)

Figure 5.1. Socket Weld Full Coupling.

automatic methods use a repetitive process, ideally suited to the fabrication shop, where the conditions can be controlled.

Manual	Semi-Automatic	Automatic
Metal arc Oxyacetylene TIG (argon arc) Flux-cored arc Gas-shield flux-cored arc	Metal inert gas (MIG)	Tungsten inert gas (TIG)

Metal Arc

Metal-arc welding also is known as *stick welding* and, in the United States, as *shielded metal arc welding* (SMAW). This process requires striking an arc between a consumable metal rod (electrode) and the

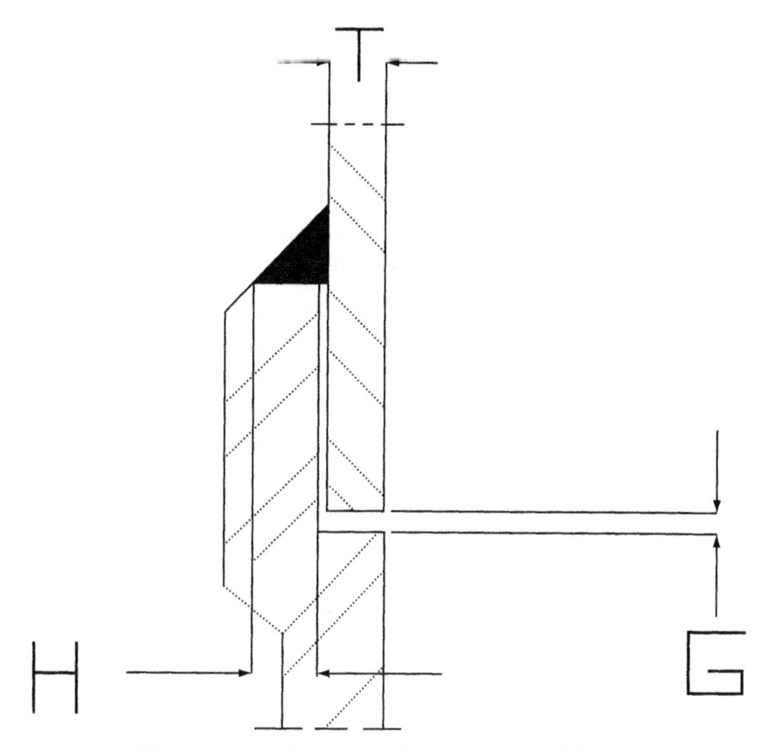

Figure 5.2. Root Gap For Socket Weld Joint.

parent metal, which is the workpiece and the two pieces of metal to be welded. Applied heat melts some of the parent metal and part of the electrode. The weld metal therefore is a mixture of the parent metal and the electrode metal. To avoid the formation of oxides that weaken the weld, the electrodes are coated, which forms a slag that protects the weld from atmospheric contamination during postweld cooling.

This electrode coating can also contain deoxidizing agents, which are deposited into the molten pool to add further protection from oxygen in the air.

Metal arc welding is used widely in the fabrication of smaller fillet welds, because it is a simple and cheap method. It is not suitable for butt welds and larger fillet welds, which require several successive passes, as the trapped slag causes a problem with this process. This form of welding requires a power source, a consumable electrode in a holder, and a struck arc.

Also, the sustained intensity of the arc makes it difficult to prevent "burning through" when welding thinner sections of steel, under 1.6 mm.

For sections below 1.6 mm, TIG welding is the preferred option, although oxyacetylene welding is used on occasion. The slag created during this process must be chipped off the weld bead after welding.

The following table summarizes the benefits and limitations of this method.

Benefits	**Limitations**
Simple equipment	Slag creation
Flux regulation	Burn through on thin sections
Lower sensitivity to wind and drafts	
All positions possible	

Combined welding methods are sometimes used. When butt welding small, thick-walled pipe, it can be difficult to achieve an even penetration; but satisfactory results can be obtained by using TIG welding for the first run and completing the remainder of the weld by using the cheaper metal arc method.

Oxyacetylene

In oxyacetylene welding, oxygen and acetylene are fed through a blowpipe, where the mixed gases are burned simultaneously at the tip, creating an intensely hot flame. This is used to heat and melt the edges of the workpieces and the filler rod, which is deposited in the molten pool to create the weld metal.

The filler rod is usually of the same composition as the workpiece and provides additional mass to create the joint. Flux is not usually required for oxyacetylene welding; however, if it is introduced, it can be applied as a paste on the edges of the workpieces or coated on the filler rods.

The temperature of an oxyacetylene flame is lower than an arc, which means that it can be used on thinner sections of metal; however, it also means that there could be a lack of fusion between the weld and the workpiece.

Submerged Arc Welding

Submerged arc welding (SAW) is a high-quality, very high-deposition-rate welding process. SAW uses a granular flux that forms a thick layer to prevent sparks and splatter and acts as a thermal insulator for deeper heat penetration. SAW provides high weld productivity, approximately 4–10 times as much as SMAW.

The following table lists the benefits and limitations of SAW.

Benefits	**Limitations**
Extremely high-deposition-rates	Irregular wire feed
High-quality welds	Horizontal position only
Easily automated	
Lower operator skill required	

Tungsten Inert Gas

Tungsten inert gas (TIG) welding, also known in the United States as gas tungsten arc welding (GTAW), is a high-quality welding process that requires the following:

- A power supply.
- A nonconsumable electrode (usually tungsten).
- An inert gas supply (argon/helium).
- A filler rod (similar in composition to the parent material).
- A struck arc.

The tungsten electrode is mounted centrally in a nozzle-shaped hook through which the inert gas is passed at a controlled low velocity, which effectively protects the weld area from atmospheric contamination.

The inert gas options include argon, argon + hydrogen, and argon/helium. Helium is generally added to increase the heat input, which increases the welding speed. Hydrogen results in a cleaner looking weld; however, its presence my promote porosity or hydrogen cracking.

Heat from the arc melts the edges of the two workpieces and the filler rod to create the molten pool, which after cooling forms the weld.

Because of the protective shielding of the weld area by the inert gas, a flux is not required for this process. Effective fluxes can also be corrosive, and their elimination is a great advantage when fabricating corrosion-resistant alloys (CRA).

If a filler wire is required, it is added to the weld pool separately.

The following table lists the benefits and limitations of TIG.

Benefits	Limitations
High-quality welds Can be done with or without filler Heat control Free of weld splatter Low distortion	Harder to perform than metal inert gas welding Slower deposition rates More costly

Flux-Cored and Gas-Shielded Flux-Cored Arc Welding

Flux-cored and gas-shielded flux-cored arc welding also is known as flux-covered arc welding (FCAW). As with metal inert gas welding, welding wire is fed continuously from a spool, and this method is the semi-automatic welding process.

It is similar to metal arc welding and requires the use of a consumable metal electrode with a flux core, which protects the weld metal. The use of the gas shield ensures added protection to the weld when required.

Metal Inert Gas Welding

Metal inert gas (MIG) welding, also known in the United States as gas metal arc welding (GMAW), offers high quality and a high deposition rate. The process consists of arc burning between a thin bare metal wire electrode and the workpiece. The welding zone is shielded by adding an inert gas, like argon, helium, carbon dioxide, or a mixture of gases. The arc is self-adjusting, and any variation in the arc length made by the welder produces a change in burn rate. Deoxidizers present in the electrode prevent oxidization in the weld pool, which allows multiple weld layers.

This process is similar to the TIG welding technique, except that the tungsten electrode is replaced by a consumable bare metal electrode of a material similar to the workpieces. Wire is continuously fed from a spool, and this is a semi-automatic welding process.

It requires the following:

- A power source, generator, or a rectifier to strike the arc.
- A consumable electrode (usually tungsten), with a feed motor.

- An inert gas supply (argon + helium).
- A torch or gun.

Consumable-electrode inert gas welding shares the same advantages as TIG welding, in that it does not require the addition of a flux.

There are several inert shielding gas options: argon, argon with 1–5% oxygen, argon with 3–25% CO_2, and argon with helium. CO_2 can be used in its pure form in some MIG welding processes. It can adversely affect the mechanical properties of the weld, however. Because of the higher temperature supplied by the arc, materials of thickness of 3 mm and above can be welded.

The advantage that MIG welding has over TIG welding is that the process is almost twice as quick. The following table list the benefits and limitations of MIG.

Benefits	Limitations
Can be done in all positions	Requires filler
Faster than MIG	
Less operator skill required than MIG	
Long continuous welds	
Minimal postweld cleaning required	

5. HEAT TREATMENT

Depending on the welding procedure, two additional heat treatment processes may be necessary to complete a weld that satisfies the codes requirements. Preheating requires the workpieces to have heat applied prior to the welding process. This involves heating the workpieces to a predefined temperature (see ASME B31.3) then allowing it to cool. Postweld heating may be necessary to restore the original metallurgical structure or reduce the residual stresses caused by differential cooling; in certain cases, this is mandatory in the code. Postweld heat treatment is best carried out in a furnace, which allows accurate control of the temperature, temperature gradients, and cooling rate. Sometimes, this is not possible; and welds have to be postweld heat treated in situ, which requires the use of portable heating elements.

6. NONDESTRUCTIVE EXAMINATION OF WELDS

It is essential that the completed weld should not have discontinuities or voids and that the mechanical strength of the weld is equal to or greater than the parent pipe. To reduce the possibility of failure during the hydrotest or, far worse, in-service failure, an inspection plan that applies a variety of nondestructive tests can be implemented to detect any weaknesses in the fabrication. Nondestructive examination means the assessment of a weld without damaging it physically and affecting its pressure sealing characteristics. Several methods are available, with different costs and differing levels of accuracy. Fully qualified personnel, who are in a position to interpret the results and take the appropriate action, must carry out all of these options:

- Visual (surface crack detection for all material).
- Magnetic particle examination (surface-crack detection for carbon steel and any magnetic metals).
- Dye penetrant examination (surface-crack detection for nonmagnetic stainless steels and other nonmagnetic metals).
- Radiography (surface and through the metal).
- Ultrasonic examination (surface and through the metal).

All nondestructive examination of welds must take place before hydrotesting of the piping system and painting (if necessary) or insulation (if necessary). This means that if a weld fails the examination test, the bare pipe can be repaired and retested, before the painting or insulation operation.

Different piping systems have different types and levels of inspection, depending on the service fluid, material, temperature, pressure, and location.

Visual

Visual inspection is the simplest and the cheapest method, and all welds must be subject to this basic method, using either the naked eye or a magnifying glass to confirm imperfections. All surfaces to be visually examined must be thoroughly cleaned.

This method is useful only to detect surface imperfections. If these are found, additional tests are employed to discover the extent of the flaw. Even if a weld is to be examined by more-accurate methods of inspection, it should be subjected to the basic visuals, because of their low cost. Also, if imperfections are detected visually, additional examination can be intensified around this area of concern.

Magnetic Particle Examination

Magnetic particle examination (MPE) is used to detect surface cracks on ferromagnetic materials, such as carbon steel. Some low alloys are magnetic, however ausentitic-chromium stainless steel is very weakly magnetic and therefore excluded from this type of examination; this is subjected to dye penetrant examination, which is covered later. The MPE method is very useful for detecting fine cracks that are invisible to the naked eye.

To carry out the examination, the weld under analysis is first strongly magnetized with an electromagnet, then fine particles of a magnetic material, such as iron or magnetic iron oxide, are applied to the surface. The magnetic powder is attracted to the edges of any surface cracks, making them visible to the naked eye.

Liquid Penetrant Examination

The liquid penetrant (or penetration) examination (LPE) method is used on metals considered to be nonmagnetic, such as ausentenitic-chromium stainless steel. This technique requires the surface application of a penetrating liquid containing a dye. The liquid is given time to seep into any surface flaws, and excess liquid is removed. The surface is allowed to dry, and the weld is examined. Flaws are indicated by the presence of dye, which is visible to the naked eye.

Radiography

Radiographic (RT) examination is the most useful nondestructive test, as it detects subsurface flaws invisible to the naked eye. This method originally employed X-rays, but today pipe joints can be examined using gamma-rays produced by portable radioactive isotopes.

All sources of radiation are potentially dangerous, and exposure over extended periods must be avoided. Personnel protection is often a requirement for technicians carrying out the radiography.

A film is placed on one side of the weld, and on the other side, the weld is subjected to X-rays in the direction of the film. As the X-rays pass through the weld, any imperfections on the surface and through the weld are detected by a dark shadow on the exposed film. No imperfection shows up as clear with uniform shade. The analysis of radiographic films requires considerable experience, and the defects that may be detected include cracks (surface and subsurface) and subsurface cavities caused by oxide film; lack of fusion; trapped slag, flux, or foreign material; and gas pockets (porosity).

Each radiograph must be recorded with the number of the weld to identify the exact location of the weld, and the names of the radiographer and inspector must also be listed. Radiographs are open to interpretation, and it is essential that the personnel used for this activity are suitably qualified.

Ultrasonic

Ultrasonic (UT) waves with a frequency of 500–5000 kHz are transmitted as a narrow beam toward a target. On reaching a metal surface with a flaw, the waves are reflected and returned to a suitable receiver. The time required for the return of the echo is a measure of the length of the path covered by the waves.

If used correctly, the ultrasonic method can approach the accuracy of radiography. The benefit of ultrasonic testing is that the equipment is portable; therefore, UT is useful when the weld is in an awkward location or needs to be examined on site.

7. INSPECTION LEVELS

The level of inspection needed depends on the connection's service, temperature range, pressure range, and location.

An example of the acceptable levels of inspection are usually represented either in percentages, such as 10%, or numerically, such as 1 in 10. This means that 10% of the welds in a piping system at that size or rating will be subjected to certain types of examination.

These levels of inspection are imposed on the mechanical contractor; however, in construction they can be increased if the failure rate is high or reduced if there are no or very few weld failures.

8. INSPECTION RECORDS

Inspection records must be retained to guarantee that the inspection levels imposed have been met. These records must include the following:

- Type of examination.
- Name of inspector and radiographer.
- Equipment used.
- Welding procedure specification (WPS).
- Welder's performance qualification (WPQ).

9. METALLIC FLANGED JOINTS

A number of flange standards are recognized internationally. The most commonly used are as follows:

ASME B16.5—for sizes ½–24 in. pressure ratings to 150, 300, 400, 600, 900, 1500, and 2500 lbs.

ASME B16.47—for sizes 26–60 in. and pressure ratings to 150, 300, 400, 600, 900, 1500, and 2500 lbs.

There are also British (BS), German (DIN), French (AFNOR), and various other national standards; however, the ASME codes are recognized worldwide and, for the purpose of clarity and space, they are referenced in this chapter.

A very common method of joining two lengths of pipe is by using metallic flanged connections. The piping components required to make this connection are as follows:

- Two metal flanges (carbon steel, stainless steel, cast iron, Inconel, etc.).
- One set of bolts (carbon steel, low-alloy steel, stainless steel, etc.).
- One gasket (rubber, graphite, Teflon, spiral wound, metal ring).

This joint requires two mating flange faces, which are pulled together by a set of equally spaced bolts with a gasket generally sandwiched between the two faces. In very special circumstances, a gasket is not used; however, this is very rare.

The pressure seal is made by a compressive force applied by tightening the bolts against the two flanges with the gasket trapped between the two faces. This method of connection allows the bolts to be loosened and the joint disassembled. This, therefore, is not considered a permanent joint, although in practice it could be in place for several years.

Basic Flange Types

Several methods are used to connect the flange to the piping system:

- Weld neck—requires a full-penetration butt weld to connect the flange to pipe. (see Figure 5.3).
- Socket weld—requires a socket weld (fillet weld) to connect the flange to pipe. (see Figure 5.4).
- Slip on—requires a minimum of one fillet weld, however some clients specify two fillet welds to connect the flange to the pipe (see Figure 5.5).
- Screwed—requires a male/female thread to connect the flange to the pipe (see Figure 5.6).
- Lap joint—requires two piping components for each side of the joint, one stub end, which is butt welded to the pipe, and a backing ring. The backing ring is drilled to take a set of bolts to make the pressure seal (see Figure 5.7).

Flange Facings

A flange must have a specified facing, which could be one of the following;

Flat face (FF)—a flat machined face, which requires a full-faced gasket to make the pressure seal.

Raised face (RF)—a flange with a raised step machined on the face, which requires a spiral wound gasket to make the pressure seal.

Ring-type joint (RTJ)—a flange with a circumferential groove machined into the flange face, which requires an oval or an octagonal circular ring gasket to make the pressure seal.

Tongue and groove (T&G)—two dissimilar flanges, one having a tongue machined on one face and the other flange having a mating groove machined onto the face.

The first three are the most common flange facings used in the process industry; the last type is available but rarely used.

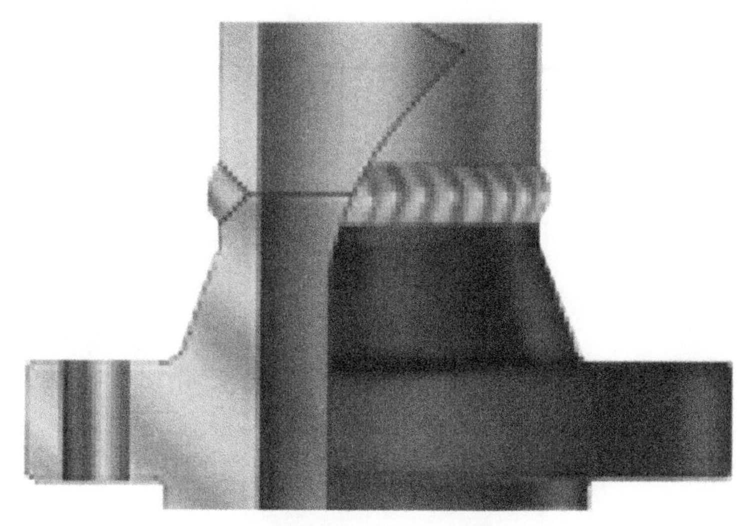

Figure 5.3. Weld Neck Flange, Raised Face.

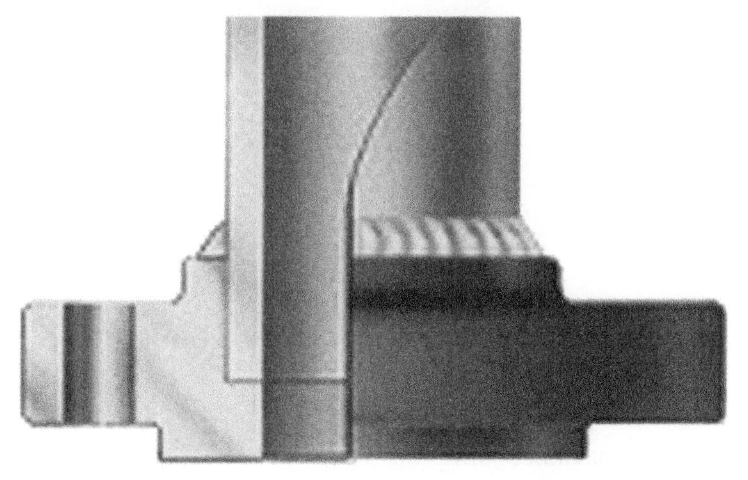

Figure 5.4. Socket Weld Flange, Raised Face.

Machining of Flange Facing

The machined faces for the flat face and the raised face flanges are supplied in various machined finishes. The grooves are concentric or

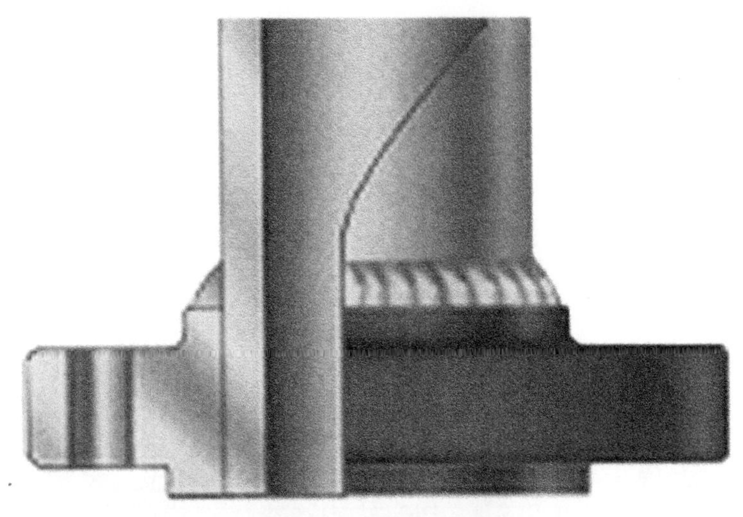

Figure 5.5. Slip On Flange, Raised Face.

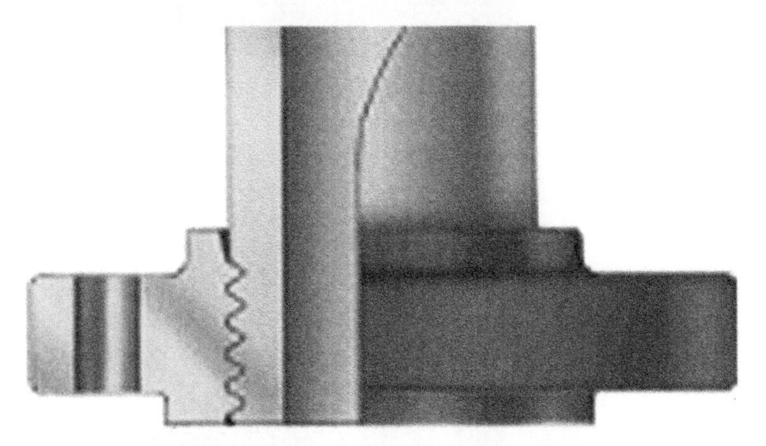

Figure 5.6. Screwed Flange, Raised Face.

phonographic, machined onto the flange face that "bites" into the gasket, provides an improved pressure seal, and prevents the gasket being squeezed out radially when the bolt loads are applied. These finishes are identified in micro inch (AARH) or micrometer (Ra). For example:

125–250 AARH (Ra 3.2–6.3)—suitable for flat faced flanges with soft cut gaskets 1.5 mm and thinner.

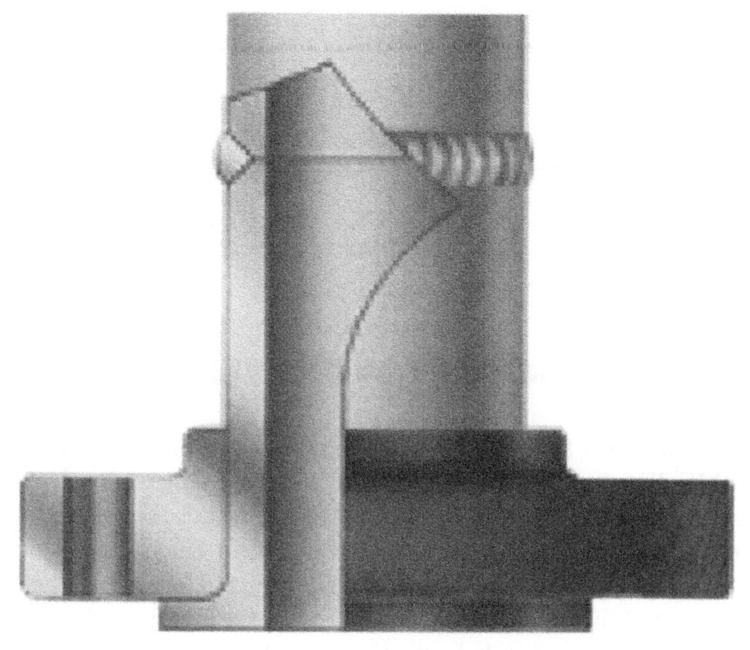

Figure 5.7. Lap Joint Flange, Raised Face.

125–500 AARH (Ra 3.2–12.5)—suitable for flat faced flanges with soft cut gaskets thicker than 1.5 mm.
125–250 AARH (Ra 3.2–6.3)—suitable for raised face flanges with spiral wound gaskets.

10. WELD-NECK FLANGE

The dimension and the design of the weld neck flange have been calculated to meet the approval of the relevant code, in this case ASME B16.5 or ASME B16.47 (series A or B). Basically, the weld neck flange comprises a flanged blade with standard drilling for bolts, based on the size and pressure rating. One side of the blade is machined to mate with another flange, and the other side has a tapered hub with a weld bevel prepared for mating to a pipe of a matching diameter.

One circumferential weld is required to connect the flange to the pipe. A butt weld is a high-integrity weld that can be inspected visually or using NDE with MPE or LPE, UT or RT, and finally subjected to a hydrostatic test.

Generally, weld neck flanges are machined from forgings, which offer more consistent grain properties through the body of the component. Despite its high cost, the weld neck flange is a preferred method of jointing for use under high-pressure, high-temperature, and cyclic loading conditions. It is also commonly used at lower pressures and low temperatures.

11. SOCKET-WELD FLANGE

The socket weld flange is designed to standard dimensions stipulated by ASME B16.5. The flange comprises a drilled flanged blade with a machined face on one side and, on the other, a female socket into which the pipe is placed. As with the weld neck, these flanges are generally made from forged steel. The flange and the pipe are joined together by a circumferential fillet weld, which is cheaper than a butt weld; however, this method of jointing is less effective and, therefore, generally used in sizes 2 in. and below at ambient and intermediate temperatures and in ASME B16.5 classes 150 lb and 300 lb. Some clients prohibit the use of socket weld flanges in process systems and limit its use to utility piping systems, air, water.

12. SCREWED FLANGES

Screwed and socket weld flanges are of very similar construction; however, instead of having a socket bored into the forging, an NPT (National Pipe Taper) thread is tapped. This allows a pipe with a matching male thread to be screwed into the flange. Because this joint does not require welding, it is both cheaper and quicker to execute than butt and socket welds.

However, it is the least efficient joint, used almost exclusively for nonhazardous utility piping systems. Because there is no weld, NDE is limited to visuals, followed by a hydrostatic test. If a leak should occur, then the system can be shut down and the screwed connection back welded, which effectively converts it into a socket weld flange.

13. SLIP-ON FLANGES

Basically, a slip-on flange is a drilled flange blade through which a hole is bored. The pipe is inserted into the flange prior to welding. The pipe is attached to the flange by one external circumferential fillet weld or two fillet welds, one external and one internal against the flange face. At smaller sizes, this method of jointing is cheaper than a weld neck; however, it lacks mechanical strength and generally is used only for utility piping classes operating at ambient temperatures and lower pressures. The relevant codes are ASME 150 lb and 300 lb

14. LAP-JOINT FLANGES

A lap-joint flange requires two piping components for each side of the flanged connection, a stub end and a loose backing flange. The loose backing flange fits over the outside diameter of the stub end, which is butt-welded to the pipe. The backing flange is not welded to the pipe, and it can be rotated, which is particularly useful when it is necessary to orientate flanges during erection.

Also, as the backing flange does not come in contact with the process fluid, it can made of a less corrosive-resistant material. For example, if the process is corrosive and requires the pipe to be of stainless steel, as in ASTM A312 TP316L, then the stub end must also be made of SS 316L; however, the backing flange can be made of the cheaper ASTM A105.

This method of jointing is not as robust as a weld neck flange but is superior to screwed, socket weld, and slip on connections; however, it is more expensive to execute, because it require a full-penetration butt weld and requires two components.

15. JOINT COMPARISON

To summarize, all the preceding methods of jointing have their advantages and disadvantages, which are shown in the following table.

Type	Characteristics	Limiting Factors
Weld neck	High-integrity joint Suitable for use at high pressures and temperatures Available at all sizes High component cost	Component more expensive than screwed, socket weld, and slip on flanges Butt weld more expensive than socket weld Heavier component
Socket weld	Intermediate-integrity joint Suitable for low and intermediate pressure and temperatures Used for smaller sizes, <4 in.(usually <2 in.) Low component cost	Not suitable for high pressure and temperatures Not available in sizes over 4 in.
Screwed	Low-integrity joint Used for utility piping systems and ambient temperature and low pressures Used for smaller sizes, <4 in. (usually <$1\frac{1}{2}$ in.)	Not suitable for process piping Not available in sizes over 4 in.
Slip on	Low-integrity joint Used for utility piping systems and ambient temperature and low pressures Available in larger sizes Low component cost	Suitable for only lower pressures and temperatures Generally restricted to utility piping systems Requires two fillet welds
Lap joint	Intermediate- to high-integrity joint Suitable for use at high pressures and temperatures Available in larger sizes	Not suitable for smaller diameters Requires two components

The choice of flange connection should be based on the following considerations.

Concerning the process media:

- Hazardous process—ASME 31.3, Category M.
- Nonhazardous process—ASME B31.3, Normal Fluid Service (NFS).
- Utility service—ASME 31.3, Category D.

For pressure conditions:

- Low pressure—ASME Class 150 and 300.
- Intermediate pressure—ASME Class 600, 900, 1500, 2500.
- High pressure—above ASME 2500 as specified in ASME B31.3, Chapter IX, K.

For joint integrity:

- High—butt weld used for weld neck, lap joint.
- Medium—fillet weld used for socket weld, slip on.
- Low—threaded used for screwed joints.

For size:

- Generally $1\frac{1}{2}$ in. and below—a choice between butt weld, socket weld, and screwed.
- Above 2 in.—butt weld is the preferred pipe to pipe weld, which excludes the use of socket weld and screwed flanges.

For component cost:

- Lap joint—most expensive (two components).
- Weld neck—expensive (one component, one butt weld).
- Socket weld—cheaper (one component, one or two fillet welds).
- Screwed and slip on—cheapest (one component, no weld).

For fabrication cost:

- Lap joint and weld neck—most expensive (one butt weld).
- Slip on—expensive (two fillet welds).
- Socket weld—cheaper (one fillet weld).
- Screwed—cheapest (one mechanical joint).

For the flange face:

- Flat face.
- Raised faced.
- Ring-type joint.

6

BOLTS AND GASKETS

Initially, I was going to create separate chapters to cover the subjects of bolts and gaskets for process piping systems; however, because they are so closely linked, I decided to place them in the same section. Bolts and the accompanying gaskets form a huge subject from an industrial perspective; however, I discuss their function only within the context of the jointing method for process piping systems. A set of bolts and gaskets are used to achieve a pressure-retaining seal when two flanges are mated together.

1. BOLTS

For ASME flanges, bolts, sometimes called *fasteners*, are used in sets, which come in multiples of four bolts, with four the minimum number. The bolt spacing is carefully calculated, taking into consideration the nominal pipe size of the flange, the design temperature, the pressure to be encountered, and the material of the flange to ensure good sealing characteristics. The outside diameter of the bolt, its length, and the thread engagement are also important factors to consider. The material of the bolt must be of sufficient strength to allow the correct bolt loads to be applied; however, a high-strength stud bolt combined with a lower strength flange results in deformation of the flange when bolt loads are applied. It is also important that the nuts have full engagement with the thread of the stud. Some companies specify that 1.5 to 3 threads must be exposed above the crown of the nut, to guarantee that full engagement has been achieved. Too many threads exposed may result in impact damage, which may make it difficult to remove the nut when maintenance is required.

Bolt Selection

A vast majority of the bolts used within the process industry come in two types: hexagonal machine bolts complete with one hexagonal nut, and more commonly, stud bolts, which have a threaded stud complete with two hexagonal nuts. Bolts come in four material groups: carbon steel, low alloy steel, stainless steel, and exotic material, such as Monel or Inconel. Each of the material groups contains a number of grades, with particular characteristics in mechanical strength and performance at low and elevated temperatures. ASME B31.3 lists all the significant bolting materials and references their yield and tensile strength through a range of temperatures. The code also specifies the minimum design temperature at which the bolt can be used, and this minimum temperature is mandatory to satisfy the code.

The two design types and the four material groups cover a vast majority of the combinations a piping material engineer is likely to experience when working on process plants.

Bolt Coating

Several types of coatings can be applied to the bolts to protect the base material from the environment local to the flange:

- None—bare bolts, with no coating, are also known as *black bolts.*
- Zinc.
- Hot-dipped galvanization.
- PTFE (polytetraflouridethylene).
- Other coatings.

Each type of coating has its advantages, and these benefits come at a price.

Hexagonal-Head Machine Bolts

This is a two-component fastener that combines a stud with an external thread and an integral head with a nut with a matching internal thread.

The term *machine bolt* refers back to when the final flanged joint between a piece of cast machinery and a piping systems, the bolt, was made from a lower-strength steel to match the mechanical strength of the cast flange of the equipment and avoid flange deformation.

Machine bolts are also used to join together two lower-strength flanged piping systems, which could be constructed of cast iron, glass-reinforced epoxy or plastic (GRE or GRP), bronze, or a similar material.

Lower-strength machine bolts can also be used to join high-strength piping flanges to lower-strength piping flanges. When joining a high-strength carbon steel flange with a raised face to a weaker flat-face cast iron flange, a full-face gasket should be used with low-strength machine bolts. This avoids the possibility of the weaker material, cast iron, bowing and breaking the effective seal and allowing the joint to leak.

Stud Bolts

Stud bolts are three-component bolts that combine a stud bolt with a thread along the length of the stud and two nuts with matching internal thread through the length external thread and two nuts with matching internal threads. Sometimes an additional nut is required for overlong bolts when controlled hydro tensioning machines are used to accurately apply loads to the bolt. This is common for bolts used in sizes of over 1 in. for larger flanges under high-pressure conditions. Hydro-tension is also specified by some clients for bolts of all sizes in toxic service. The exposed thread required for the hydro-tension machine must be covered by a third nut to protect it from mechanical impact and damage to the thread.

The stud is generally threaded for the entire bolt length, even though the center section of the thread may never come in contact with either nut.

Bolt Coatings

To afford protection from the environment, bolts can be protected by an applied coating. Some operators do not protect bolts with a coating in many environments, because they consider that these bolted joints, once they are torqued and hydro tested, will be maintained until the complete set of bolts is changed. External corrosion is negligible, and so the weight loss of material, resulting in a lower strength bolt, is no consideration.

Bolting should be changed in sets so that all the individual bolts have the same mechanical strength and loads, when applied, are uniform. Heavy corrosion in the threads of these bolts is no problem, because the threads are not be used again; nut splitters are used to crack open the nut and release the stud.

The reuse of bolts should be avoided; although, in some cases, this might not be possible. If the bolts have to be reused, it is better to use the entire old set, because all the bolts in it have similar mechanical strength. Adding several new bolts could result in unequal bolt loads, because of the differing mechanical strengths.

The most common bolt coatings are zinc, hot-dipped galvanization, and PTFE.

Lubrication of Bolts

Before loads are applied to bolts, it is essential that a lubricating coating be applied to the external thread of the bolt and the internal thread of the nut to reduce the coefficient of friction. This means that lower bolt loads are necessary to achieve an effective seal.

It is advantageous to coat the length of the bolt that will come in contact with the nut, not only the final section of engagement between the bolt and the nut. It is very important to lubricate the underside of the nut to further lower the effects of friction. This coat also gives a degree of external protection against the local environment.

Various types of lubricants are used. Each has its own characteristics, and selection should be based on the following factors:

- Lubrication—the better the lubricant, the lower the effects of friction.
- Compatibility—the lubricant must be compatible with the stud and nut and also the gasket construction; and it must not contaminate the process fluid.
- Temperature—the lubricant must be suitable at the upper and lower temperatures of the process fluid.

2. GASKETS

A gasket is a sealing component placed between flanges to create a static seal between the two stationary flanges of a mechanical assembly and maintain that seal under all design and operating conditions, which may vary depending on changes in pressure and temperature during the lifetime of the "flange."

Initially, the type of gasket chosen is based on the following criteria:

- Temperature of the media.
- Pressure of the media.
- Corrosive nature of the media.
- Viscosity of the media.
- Chemical resistance to the process media.
- Compressibility.
- Creep resistance.
- Ability to corrode the mating flange.

Secondary criteria to consider include ease of handling, availability, and cost. Note that I have placed the price of the flange last, because if the design criteria and the logistics cannot be satisfied, then the price is incidental.

Temperature. The design and material of a gasket must have the mechanical strength and characteristics to meet the full design temperature range of the media contained in the piping system.

Pressure. The design and the material chosen must have the mechanical strength and characteristics to meet the full design pressure range of the media contained in the piping system.

Corrosion Resistance. The gasket must be capable of resisting chemical attack by the media being transported and the external environment, especially in the case of an undersea location.

Viscosity. The viscosity of the media should be considered. Some fluids are considered more "searching," and this must be considered during gasket selection.

Compressibility. The gasket selected must have compression characteristics to allow the seal to be effective when the appropriate bolts are applied.

Robustness. The gasket design and mechanical strength must be capable of withstanding all movement in the presence of temperature and pressure cycles that may occur during commissioning and operation.

Creep. The gasket should not creep or flow under the influence of pressure, temperature, and applied bolt loads. Creep allows the bolts to relax and therefore reduces the gasket sealing surface area and promotes a leakage.

Handleability. The gasket chosen must be easy to handle when transporting from the warehouse to the point of erection. Large, soft gaskets and spiral-wound gaskets are liable to be damaged during this phase, and spiral-wound gaskets should be well protected. Metal ring-joint gaskets have their own integral strength, so damage is less likely to happen, although care must be taken to ensure that radial damage across the sealing face of the ring does not occur.

Availability. It is pointless to select a gasket design with materials of construction that are rare and difficult to find. There is no benefit from getting the wrong gasket to the job site quickly, if it fails in service.

Cost. A cheap gasket that does not meet the design criteria should not be considered.

To conclude, to select a gasket the following have to be considered: design, materials of construction, delivery, and cost.

Types of Gaskets

For flanges designed to ASME B16.5, B16.47 (series A or B), or API 6A, materials fall into three fundamental types:

- Nonmetallic—flat rubber, elastomers, graphite, Teflon, and the like.
- Semi-metallic or composite—spiral wound, jacketed, Kamprofile (stainless steel/graphite, Inconel/graphite, and so forth).
- Metallic-ring type—soft iron, stainless steel, Monel, and the like.

Nonmetallic Materials. Numerous nonmetallic materials are used for gaskets: card, cork, elastomers, graphite. The first two, card and cork, are rare in the process industry. A vast majority of the nonmetallic gasket materials used in the process industry come from the elastomer and graphite families. These are commonly called *soft gaskets* or *cut gaskets*, because they are cut from sheet. They are easily compressed with low bolt loads. Generally, these gaskets are used for low-pressure ASME 150 and ASME 300 class and occasionally for medium-pressure ASME 600 class. Depending on the type of elastomer, they can be used for temperatures up to 392°F (200°C). Graphite gaskets are suitable for temperatures up to 1022°F (550°C).

Soft gaskets are generally the cheapest type of gasket.

Rubber and Elastomers. This group includes, natural rubber and the many synthetic grades of elastomer, like neoprene, nitrile, butyl, ethylene propylenediene, styrene butadiene, and Viton. Each elastomer has its own mechanical characteristics and resistance to process media. When used independently, these elastomeric materials are selected for media at lower design temperatures, 392°F (200°C) maximum and low pressures, ASME 150 and 300 class. Elastomers are best suited to transport noncorrosive hydrocarbons and for utility services.

The design temperature ranges for natural rubber and elastomers are as follows:

- Natural rubber (NR)—design temperature –30°C to 70°C.
- Neoprene (CR)—suitable for use with hydrocarbons, moderate acids or alkalis, saline solutions; temperature range –40°C to 100°C.
- Nitrile (NBR)—design temperature range –40°C to 100°C.
- Butyl (IIR)—design temperature range –30°C to 120°C.
- Ethylene propylenediene (EPDM)—design temperature range –40°C to 150°C.
- Styrene butadiene (SBR)—design temperature range –60°C to 80°C.
- Viton (FPM)—15°C to 200°C.

These are conservative figures and the limits of the temperature ranges can be slightly extended. Most clients have data sheets that stipulate the upper and lower temperatures at which these elastomers can be used within their projects.

Graphite. Graphite and carbon-based materials can be used at higher temperatures than elastomers, up to 500°C, but as with elastomers, they are better suited to lower pressure, ASME 150 and 300 class, when used individually without reinforcement. The graphite family can be used with a variety of media such as chemicals, hydrocarbons, and steam.

PTFE. Although PTFE's operating limitations are only marginally higher than that of rubber and the elastomer family, it is virtually chemically inert and it can be used for most services.

Semi-Metallic or Composite Materials. Composite gaskets are made from different materials to satisfy temperature and pressure requirements and make the complete assembly more robust. For example, a spiral-wound gasket can have four separate elements:

- Metal windings—to hold the filler.
- Filler—the sealing medium.
- Outer ring—to hold the gasket in the bolt circle.
- Inner ring—to prevent the windings collapsing into the process fluid.

When it is necessary to select gaskets to seal flanged joints designed for pressures ASME 600 class and above, the gasket must be more robust; therefore, it is constructed of nonmetallic materials suitable for higher temperatures and metallic materials for mechanical strength. The most common nonmetallic-metallic combination is the spiral-wound gasket that combines stainless steel windings with a graphite filler material and inner and outer rings. Gaskets of this construction can be used at temperature up to 500°C and for pressure classes from ASME 150 to ASME 2500.

Other semi-metallic gaskets include Kamprofile, metal jacketed, and reinforced or tanged, each with its own specific sealing characteristics and associated cost difference.

Because of the complexity of their construction, semi-metallic gaskets are more expensive than soft gaskets.

Ring Gaskets. Metallic ring gaskets are used for higher process pressures and commonly used for ASME 900 class and above. These gaskets are very robust. The ring fits in a groove machined out of the flange face, and both mating flanges are identical. The cross section of the ring can be either oval or octagonal.

These rings are available in a variety of materials to suit the process media and the design pressures and temperatures: soft iron, stainless steel (various grades), and exotic metals, such as Monel, Inconel, or titanium.

The ring is contained within the groove. It deforms at the base when bolt loads are applied, resulting in a very effective seal. Although slight deformation takes place during this process, the actual flange faces never touch when the bolts are tightened. A hardness differential between the gasket (softer) and the surface of the groove (harder) ensures that the gasket deforms and not the face of the flange.

Generally, because of the mass of metal and precision machining, the ring type joint gasket is the most expensive of the three options.

Flange Surfaces Machined Finishes

Flanges can be supplied with a variety of machined finishes to complement the type of gasket being used for that particular pressure and temperature.

The finished surfaces are created by machining concentric grooves or "gramophone/spiral" grooves onto the surface of the flange, and they can vary in depth. The depth of the groove is measured in microns or AARH.

3. BOLTS AND GASKETS

The Seal

To complete a pressure seal, the following components are required:

- Two flanges.
- One gasket.
- A set of bolts.

A seal is completed by compressing the gasket material with applied bolt loads between the two flange faces. This causes the face of the gasket to flow into the imperfections on the gasket seating surfaces, so that contact is made between the gasket and the two flange faces, preventing the escape of the confined fluid.

Compression (Figure 6.1) is by far the most common method of effecting a seal on a flange joint, and the compression force is normally applied by the bolts.

Bolting Procedure

For an effective seal, not only must the correct gasket design and material be chosen, but the method of installation must be correct. This process installation includes the following:

- Lubrication of the bolts.
- The bolting sequence.
- Application of bolt loads.

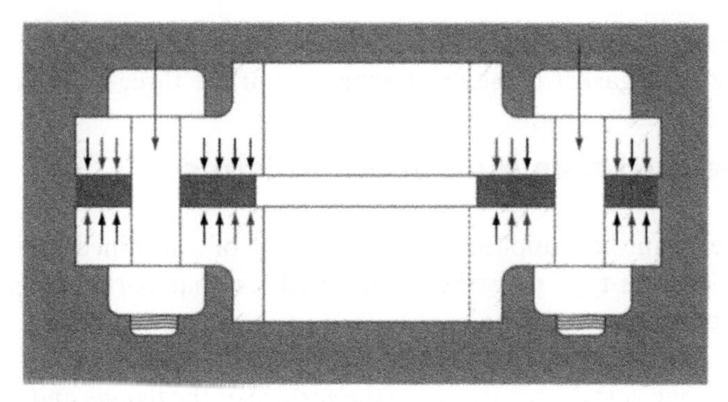

Figure 6.1. Compression Effecting a Seal on a Flange.

Lubrication of the Bolts. A recommended lubricant must be used to reduce the friction between the stud, the nut, and the back of the flange, which is why it is important that the underside of the nut also be lubricated.

Care should be taken that the lubricant does not reach the gasket, because of possible contamination; this is not essential for metal ring-type gaskets, but it is important for soft and semi-metallic gaskets.

Bolting Sequence. The gasket must be compressed and seated evenly, so it is essential that a procedure is followed to achieve this uniformity. There are a number bolting procedures, each with slightly different bolt torque percentages and numbers of steps, but this is the most common:

Step 1. Hand tight.
Step 2. Apply 30% of the final torque.
Step 3. Apply 60% of the final torque.
Step 4. Apply the final torque.

To evenly distribute the bolt loads, after a bolt has been tightened to one of these steps, the next bolt to be tightened must be the one directly opposite, 180° away. A final pass may be made in a clockwise direction, to check that all bolts have been correctly tightened and no step has been missed. This is particularly important for large flanges, which could have 32, 36, or more stud bolts. The sequence for a four-bolt flange would be north, south, east, west. To conclude, well-lubricated bolts, a closely followed bolting procedure, coupled with this bolting sequence, results in a well-seated gasket that comfortably holds the hydro-test pressure, which could be up to 1.5 times the design pressure.

It is pointless to select the correct materials of construction, bolts and gaskets, if they are assembled incorrectly. The flange joint will fail either during the hydro test or, worse still, in service, which will result in a system shutdown or a very costly plant shutdown. The piping material engineer is responsible for writing the "Fabrication and Erection Specification," which must mention lubrication of bolts, bolting procedure, and the sequence in the narrative. This is a guide for the contractor, who if experienced will carry this out automatically, but it is the responsibility of the piping material engineer to detail this as part of his or her responsibilities.

7

VALVES

Valves are essential components of a piping system, and they allow the process fluid to be controlled and directed on its journey through the process plant. They are expensive engineered items, and it is important that the correct valve is specified for the function and that it is constructed of the correct material for the process fluid.

There are two methods of operating a valve: manually, with a handwheel, lever, wrench, or actuator; or through automatically controlled valves. The piping material engineer is responsible for specifying and requisitioning valves of the first group; however, the second group are tagged valves, and they are generally the responsibility of the instrument engineer.

Table 7.1 shows the various types of valves available for specific functions. This is a basic guide. Once the type of valve has been selected, the design can be finalized, design codes referenced, and the materials of construction chosen.

1. RESPONSIBILITY FOR VALVE SELECTION

The piping material engineer is responsible for the valves selected for both the process and utility piping systems in a process facility. The valves selected must meet the fluid design conditions, handling the pressure and temperature limits and the corrosion characteristics of the fluid.

Table 7.1. Valve Selection Based on the Fluid Phase and Type and the Function the Valve Performs

Nature of Fluid	Valve Function	Type of Disc
Liquid Phase		
Neutral (water, oil, etc.)	On/off	Gate Rotary ball Plug Diaphragm Butterfly Plug gate
	Regulating	Globe Butterfly Plug gate Diaphragm Needle
Corrosive (acid, alkaline)	On/off	Gate Plug gate Rotary ball Plug Diaphragm Butterfly
	Regulating	Globe Diaphragm Butterfly Plug gate
Hygienic (food, beverages, drugs)	On/Off	Butterfly Diaphragm
	Regulating	Butterfly Diaphragm Squeeze Pinch
Slurry	On/off	Rotary ball Butterfly Diaphragm Plug Pinch Squeeze
	Regulating	Butterfly Diaphragm Squeeze Pinch Gate
Fibrous suspensions	On/off, regulating	Gate Diaphragm Squeeze Pinch

Gas Phase		
Neutral (air, steam)	On/off	Gate Globe Rotary ball Plug Diaphragm
	Regulating	Globe Needle Butterfly Diaphragm Gate
Corrosive (acid vapors, chlorine)	On/off	Butterfly Rotary ball Diaphragm Plug
	Regulating	Butterfly Globe Needle Diaphragm
Vacuum	On/off	Gate Globe Rotary ball Butterfly
Solid (powder) Phase		
Abrasive powder (silica)	On/off, regulating	Pinch Squeeze Spiral sock
Lubricating powder (graphite, talcum)	On/off, regulating	Pinch Gate Spiral sock Squeeze

2. VALVE DATA SHEETS

It is common for the project process engineers to define the function of the valve, and the piping material engineers to specify the valve on a valve data sheet (VDS), which gives complete details on the design codes, design conditions, materials of construction, testing and inspection, coating and painting, and all other requirements for the valve.

The VDS is the passport for the valve; and it must be retained as a permanent record by the client when the plant is mechanically complete, commissioned, and handed over to Operations. A maintenance manual is supplied with the valve that gives instructions on how to repair and replace components, such as seats, seals, and handwheels.

3. THE FUNCTION OF A VALVE

Valves, which come in a variety of types, are selected to perform a specific function:

- On or off—gate valve, ball valve, plug valve.
- Throttling, fluid control—globe valve (sizes to approx. 16 in.), butterfly valve (for larger sizes).
- Prevent flow reversal—swing check valve, wafer check valve, piston check valve.
- Speed of operation—multiturn (gate, globe, or quarter turn), ball, plug.
- Very special service—pinch valves, thru conduit, nonslam check.

4. MATERIALS OF CONSTRUCTION

Valves are made up of numerous components, each one having a specific function and constructed in a material suitable for that function. These components generally are made of metallic or nonmetallic materials.

Metallic Components

Metallic components are categorized by the requirements of the valve's task:

1. Pressure-containing component and with areas in contact with the process material, such as the body and bonnet.
2. Non-pressure-containing but inside a pressure containing envelope and with areas in contact with the process material, such as the stem or seat.
3. Outside a pressure containing envelope, such as a handwheel, bolts, nameplate, support, cover plate.

All components in the first group must have both the mechanical strength to cope with the design conditions and the correct material chemical composition to handle the corrosion characteristics of the process fluid.

If the component falls into the second group, then pressure containment is not an issue, but the material chosen must have the mechanical strength for its chosen function. For example, a stem material must be able to support the torque applied to open and close the valve without failure. Also, as a wetted component (in contact with the process material), the stem must have corrosion resistance characteristics for the process fluid.

The components in the third group are not exposed to the process fluid, so corrosion resistance is not consideration. They must be of sufficient strength to be functional. Bolts must be of sufficient strength to seat the gasket when bolt loads are applied and create an effective seal. Handwheels must be constructed of a robust material to ensure that they do not crack and fail when being operated.

The pressure containing envelope is that volume exposed to the full-operating conditions of the fluid temperature and pressure. Wetted describes a component directly exposed to the process fluid, either fully or partially.

Environmental conditions must be considered, and these components may require an additional coating, as is the case of valves in marine locations, which may require a coating of primer or primer and painting.

Nonmetallic Components

The pressure- and non-pressure-containing components must satisfy the operating requirements for the valve and not degenerate while in contact with the process fluid. They must also have a level of mechanical strength robust enough to suit the purpose for which they are designed:

- Primary seals—pressure containing and wetted.
- Secondary seals—pressure retaining and partially wetted.
- Soft seats—pressure containing and wetted.
- Gaskets—pressure containing and partially wetted.

All nonmetallic components form some sort of seal, either a primary seal (the first seal, and directly in contact with the process fluid and exposed to full design conditions, pressure, and temperature) or a secondary seal (any seal after the primary seal and not in direct contact with the process fluid and full design conditions, pressure, and temperature).

All the relevant valve design standards—ASME, BS, API—reference the numerous components included in the various types of valves.

It is essential that all the valve components are suitable for the process fluid and the design conditions. A chain is as strong as its weakest link, so it is pointless to select suitable material for all but one component, because this inferior part may lead to the total failure of the valve and costly maintenance.

5. ACCEPTABLE ALTERNATIVES

The piping material engineer may be offered several material alternatives; however, it is essential that the chosen component be equal to or better than that specified in the valve data sheet.

Deviation Requests

Deviations from the specification are considered before placement of order; these usually are alternatives offered by the vendor. The deviation could be in the design, the materials of construction, or the level of inspection and certification. These deviations must be agreed to prior to placement of the purchase order. Generally, the valve data sheet is modified to reflect the changes, and this becomes the standard.

Concession Requests

Occasionally, after the purchase order has been placed, the manufacturer may experience problems with subsuppliers, and the material for certain components may not be available.

The manufacturer then submits a concession request to the purchaser, offering alternative materials. The material offered must meet the project specifications. If the alternative is acceptable, then a concession request is signed and approved by the purchaser.

Relaxations

There are some exceptions, when relaxations to the specification are granted by the client. A vendor may offer a material that does not meet a very stringent client requirement but meets all the relevant international design codes; then the suggestion can be seriously considered. The alternative can be presented to the client, along with the necessary support documentation. It is the client's option to relax the corporate specification and accept the alternative offered for material availability, commercial, or delivery reasons.

6. INSPECTION AND TESTING

All valves must be constructed to a particular design code, and this specification references the standard testing to which the valve is subjected to after assembly. The level of inspection and testing can be increased at the "end users" request. This usually depends on the criticality of the valve; a large valve under a high-pressure, high-temperature, toxic fluid service is likely to be subjected to additional testing and inspection. A small bore valve in a low-pressure utility service is likely to be subjected to the lowest level of inspection.

A piping material engineer must set the priorities and ensure that, if necessary, critical valves are given sufficient attention and low-pressure process and utility valves are spot checked. Most major EPC (Engineering, Procurement, and Construction) companies have corporate inspection and testing plans that can be made project-specific to satisfy client demands.

7. CERTIFICATION

To meet quality assurance and quality control (QA/QC) requirements, all valves must carry certification to satisfy the end user or client and guarantee that the plant or project is an insurable asset. The plant owner must satisfy the insurance company that the plant has been constructed to recognized industry codes, for material and methods of fabrication.

8. BASIC MATERIALS OF CONSTRUCTION

All valves are made up of several components, each with a specific function.

Gate Valves

The principle components that make up a gate valve are listed with their requirements:

- Body—pressure containing, wetted.
- Bonnet—pressure containing, wetted.
- Bolts—mechanical strength.
- Gate—pressure retaining, wetted.
- Seats—pressure retaining, wetted.
- Stem—mechanical strength to deliver torque, wetted.
- Handwheel—mechanical strength to deliver torque.

Globe Valves

The principle components that make up a globe valve are listed with their requirements:

- Body—pressure containing, wetted.
- Bonnet—pressure containing, wetted.
- Bolts—mechanical strength.
- Disc—pressure retaining, wetted.
- Seats—pressure retaining, wetted.
- Stem—mechanical strength to deliver torque, wetted.
- Handwheel—mechanical strength to deliver torque.

Ball Valve

The principle components that make up a ball valve are listed with their requirements:

- Body—pressure containing, wetted.
- Body bolts—mechanical strength.

- Ball—pressure retaining, wetted.
- Seats—pressure retaining, wetted.
- Seals—pressure retaining, wetted.
- Stem—mechanical strength to rotate the ball, wetted.
- Lever—mechanical strength to rotate the stem.

Swing Check Valve

The principle components that make up a check valve are listed with their requirements:

- Body—pressure containing, wetted.
- Bolted blind—pressure containing, wetted.
- Bolts—mechanical strength to seat the gasket.
- Gasket—pressure containing, part wetted.
- Clapper—pressure retaining, wetted.
- Seats—pressure retaining, wetted.

Dual-Plate Check Valve

The principle components that make up a dual-plate check valve are listed with their requirements:

- Body—pressure containing, wetted.
- Plates—pressure retaining, wetted.
- Stem—mechanical strength to hold plates, wetted.
- Seats—pressure retaining, wetted.

Plug Valve

The principle components that make up a plug valve are listed with their requirements:

- Body—pressure containing, wetted.
- Plug—pressure retaining, wetted.
- Seats—pressure retaining, wetted.
- Stem—mechanical strength to rotate the stem.

9. TYPICAL GENERAL ARRANGEMENTS OF VALVES

Ball Valve—Split Body, Floating Ball

Regardless of the materials for construction of a split body, floating ball valve, all such valves have the same principal components (see Figure 7.1). The body in a split body design can be made in two pieces or three pieces. Both designs allow the ball valve to be removed from the line and repaired locally or, ideally, in a workshop. The three-piece version is more expensive but easier to maintain, because you can work on both sides of the ball.

The floating ball design means that the ball is suspended from the stem and rests on the soft seats. It is used for smaller sizes and lower- and medium-pressure classes. As the line size increases, the mass of the ball increases and reaches a weight at which it must be supported from below with a trunnion (see Figure 7.1).

The valve is available with a reduced port (usually one size down from the line size, e.g., 8 × 6 in.) or a full port (the port and line size are the same, e.g., 8 × 8 in.)

Components. These are the principal components of the valve:

- Body.
- Body bolts.
- Ball.
- Seats.
- Seals.
- Stem.
- Lever.

Added to these components and necessary to complete the construction of the valve and make it functional are the following:

- Stop pin.
- Packing gland.
- Gland nut.
- Spring washer.
- Thrust bearing.

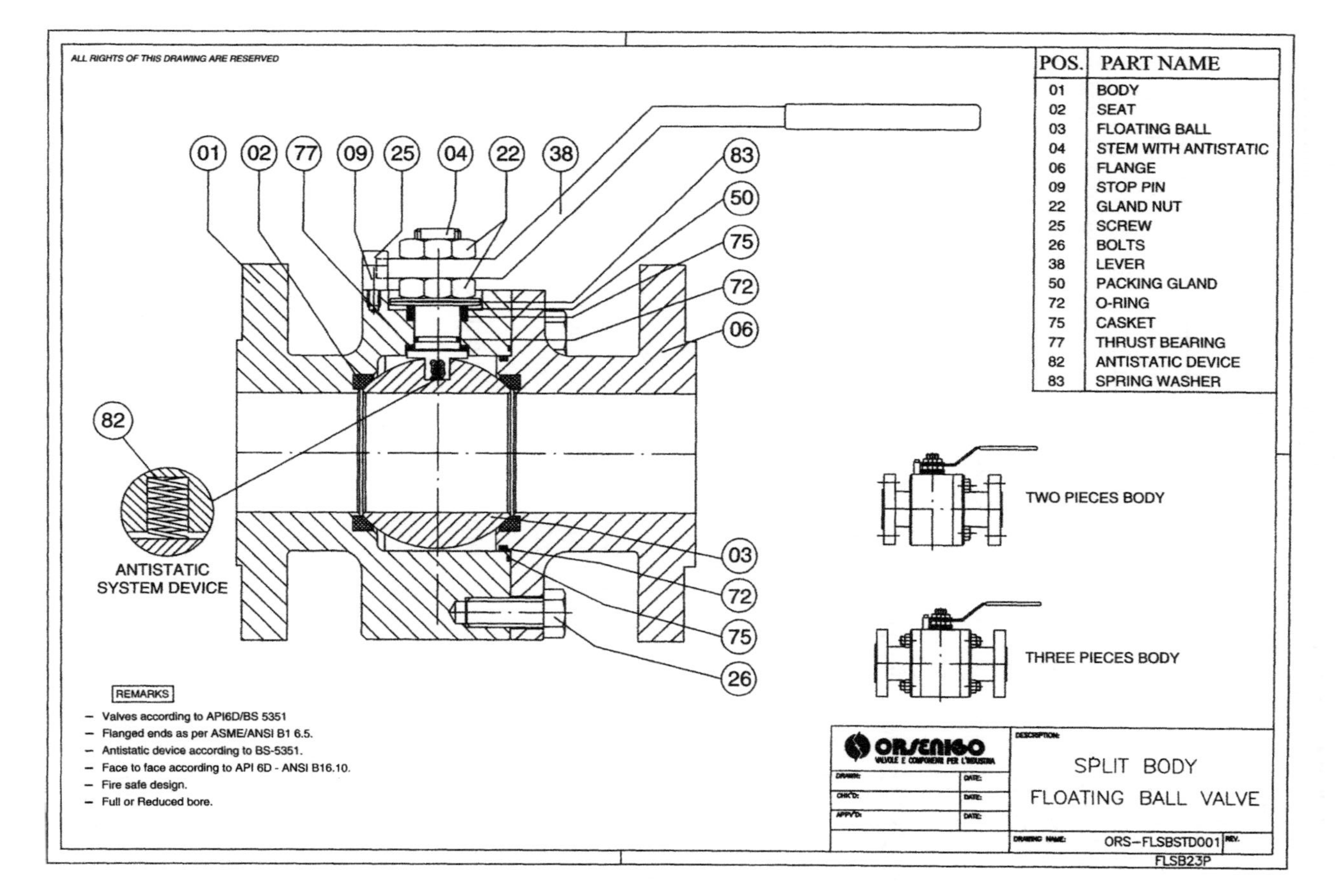

Figure 7.1. Split Body, Floating Ball Valve. (Courtesy of Orsenigo, Italy)

An antistatic device is also included to prevent a static charge as the metal ball travels over the soft seats, which could be made of PTFE.

Depending on the process conditions, some of the materials could change; others remain the same.

Design Codes. This particular valve is designed to a combination of API6D and BS 5351 specifications. The flanged ends are designed and drilled to the specifications of ASME B16.5. The antistatic device is according to BS 5361. The face-to-face dimensions are from API6D and ASME B16.10.

It is fire safe to an undefined code.

Ball Valve—Split Body, Trunnion Mounted

The valve in Figure 7.2 also has a split body for the reasons mentioned previously and is available in reduced and full port versions. However, the ball is trunnion mounted.

Trunnion-mounted valves are specified when the mass of the ball is such that it requires additional support at its base or for service at higher pressure ratings, when it is essential that the construction of the valve be more robust and the ball maintained in a fixed position when the valve is fully closed and not forced up hard against the soft seats, which risks squeezing them out of their retaining seat ring.

Components. These are the principal components of the valve:

- Body.
- Body bolts.
- Ball.
- Seats.
- Seals.
- Stem.
- Lever.

Added to these components and necessary to complete the construction of the valve and make it functional are the following:

- Stop pin.
- Packing gland.

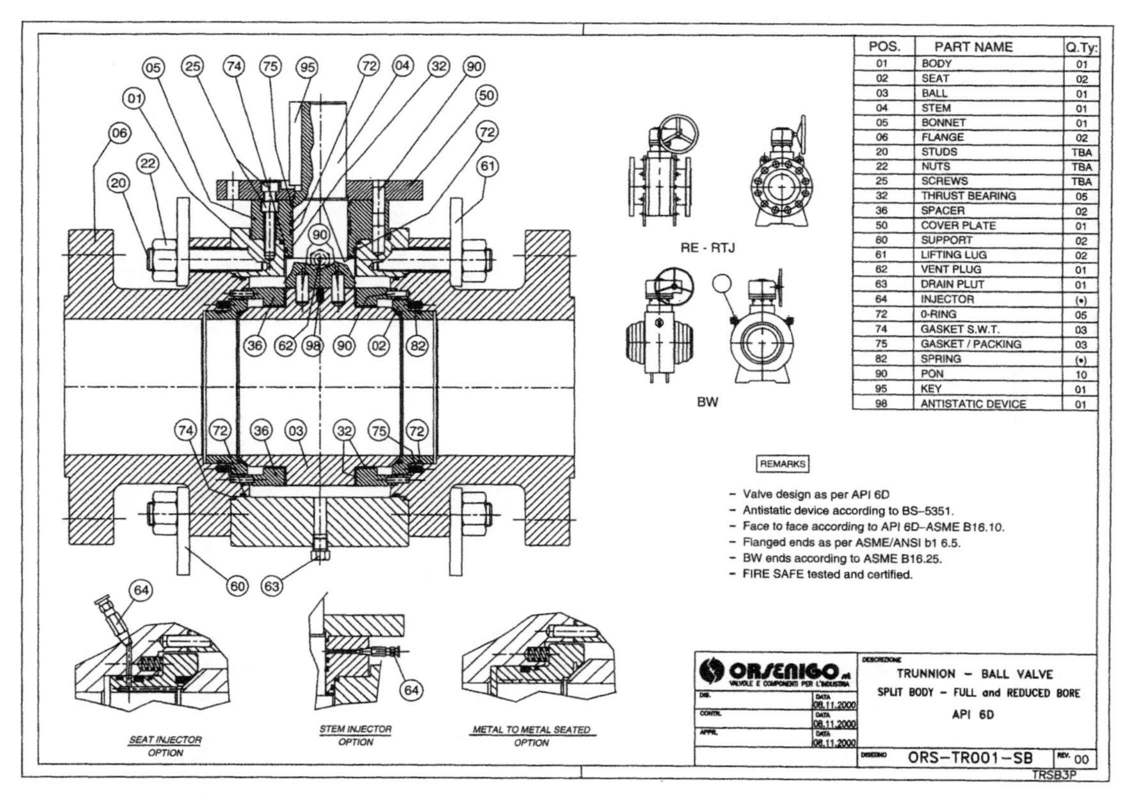

Figure 7.2. Trunnion Ball Valve, Split Body, Full abd Reduced Bore, API6D. (Courtesy of Orsenigo, Italy)

- Gland nut.
- Spring washer.
- Thrust bearing.

Because of its more complex construction and use in larger sizes and at higher temperatures than the less-complicated floating ball valve, several additional features are included, such as the following:

- Drain plug—to drain trapped fluid from the cavity between the two seats.
- Injector—a point to add scalant.
- Lifting lug—for installation.
- Support—to reduce the loads on the two mating flanges.

Three options are available:

- Seat injector—a point to add sealant when required to the seats.
- Stem injector—a point to add sealant when required to the stem.
- Metal seats—for service in an erosive environment or one that operates at temperatures above the limits of the soft seats.

This valve is designed to a combination of API 6D specifications for trunnion-mounted ball valves. The flanged ends are designed and drilled to ASME B16.5 specifications, but this valve is also available with butt weld ends to ASME B16.25 specifications. The antistatic device is according to BS 5361. The face-to-face dimensions are to API 6D and ASME B16.10 specifications.

It is fire safe to an undefined code.

Ball Valve—Split Body, Trunnion Mounted, for Cryogenic Service (below −50°F)

The valve in Figure 7.3 is very similar in construction to the pervious valve, but it has an extended stem that distances the body of the valve, which is at subzero temperatures, from the operating device, which is either a wrench or an actuator. These are to protect personnel operating the valve against frost burns from effects of the cryogenic service.

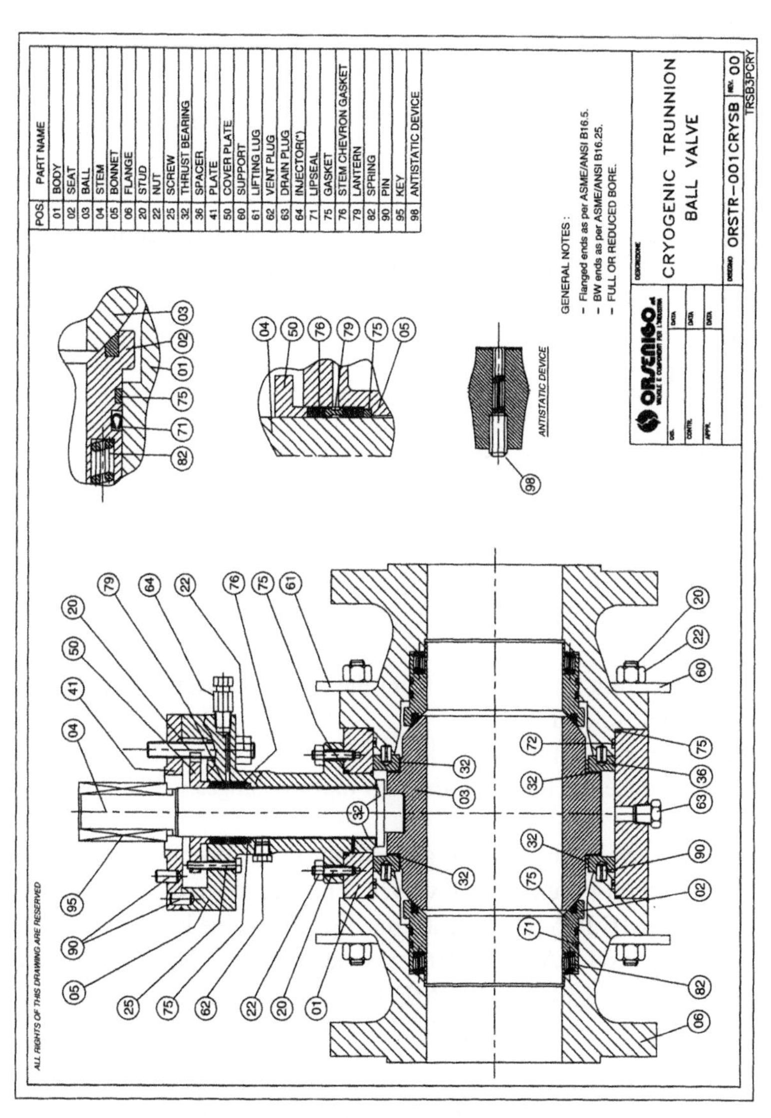

Figure 7.3. Cryogenic Trunnion Ball Valve, Split Body. (Courtesy of Orsenigo, Italy)

Components. These are the principal components of the valve:

- Body.
- Body bolts.
- Ball.
- Seats.
- Seals.
- Stem.
- Lever.

Added to these components and necessary to complete the construction of the valve and make it functional are the following:

- Stop pin.
- Packing gland.
- Gland nut.
- Spring washer.
- Thrust bearing.

Because of its more complex construction and use in larger sizes and at higher temperatures than the less-complicated floating ball valve, several additional features are included, such as the following:

- Drain plug—to drain trapped fluid from the cavity between the two seats.
- Injector—a point to add sealant.
- Lifting lug—for installation.
- Support—to reduce the loads on the two mating flanges.

This particular valve is designed to a combination of specifications from API 6D for trunnion-mounted ball valves. The flanged ends are designed and drilled to ASME B16.5 specifications, but this valve is also available with butt-weld ends to ASME B16.25 specifications. The face-to-face dimensions are to API 6D and ASME B16.10 specifications.

It is fire safe to an undefined code.

Ball Valve—Top Entry, Trunnion Mounted, Metal-to-Metal Seat

The valve in Figure 7.4 is of a top entry design, which means that it can be repaired in situ, without removing it from the line. This allows the

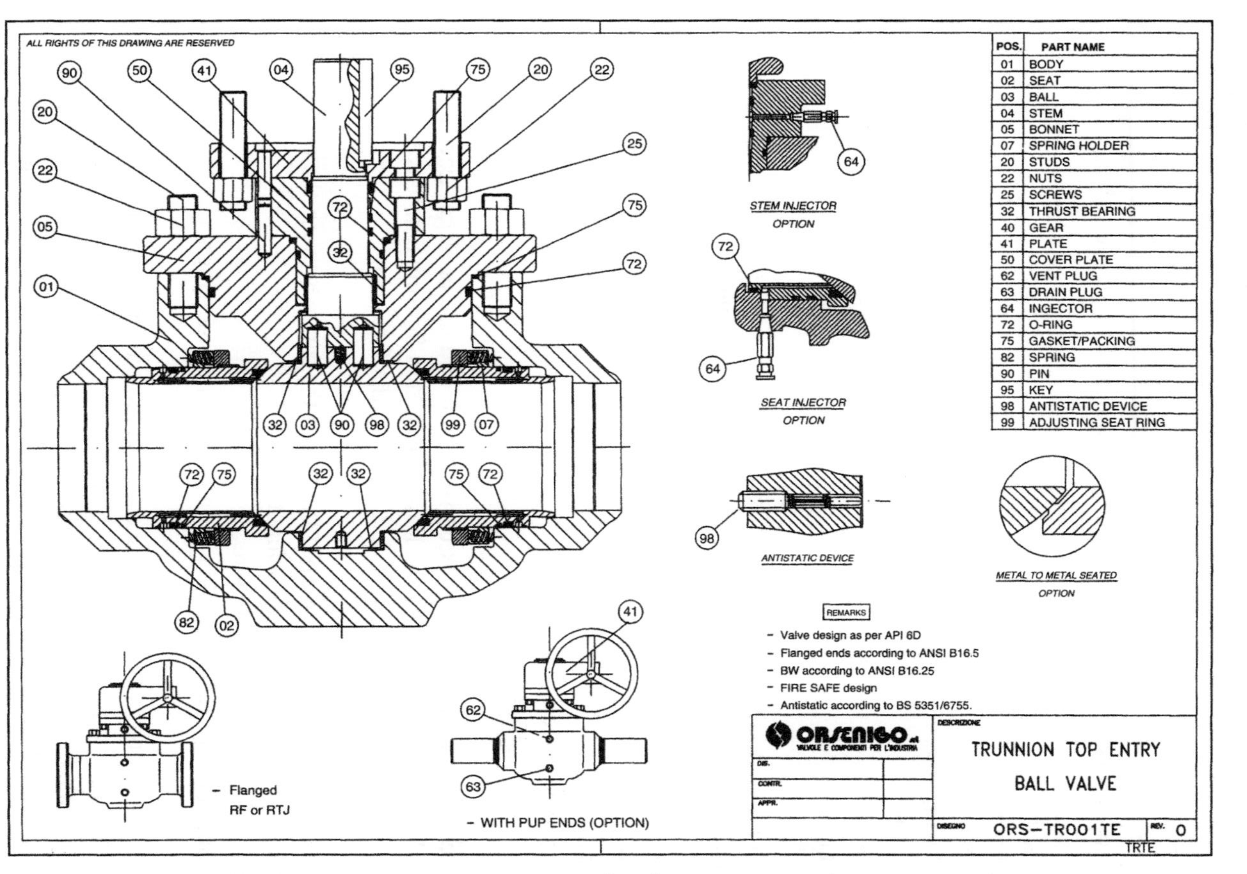

Figure 7.4. Trunnion Top Entry Ball Valve. (Courtesy of Orsenigo, Italy)

valve to be welded into the line, because it is not necessary to take it to the workshop. The metal seating allows the valve to be used in abrasive service which would scour soft seats or when temperatures exceed those allowed for soft seats.

Components. These are the principal components of the valve:

- Body.
- Body bolts.
- Ball.
- Metal seats.
- Seals.
- Stem.
- Lever.

Added to these components and necessary to complete the construction of the valve and make it functional are the following:

- Stop pin.
- Packing gland.
- Gland nut.
- Spring washer.
- Thrust bearing.

Because of its more complex construction and use in larger sizes and at higher temperatures than the less-complicated floating ball valve, several additional features are included, such as the following:

- Drain plug—to drain trapped fluid from the cavity between the two seats.
- Injector—a point to add sealant.
- Lifting lug—for installation.
- Support—to reduce the loads on the two mating flanges.

This particular valve is designed to a combination of specifications from API 6D for trunnion-mounted ball valves. The flanged ends are designed and drilled to ASME B16.5 specifications, but this valve is also available with butt-weld ends to ASME B16.25 specifications. The face to face dimensions are to API 6D and ASME B16.10 specifications.

It is fire safe to an undefined code.

Ball Valve—Top Entry, Floating Ball, Soft Seated or Metal Seated

The valve in Figure 7.5 is of a top entry design, and because of its smaller size, it has a floating ball with soft seats or the option of metal seats. The end can be either a socket weld, threaded NPT, or a combination of the two. Also, being top entry, it can be repaired in situ.

Components. These are the principal components of the valve:

- Body.
- Studs.
- Ball.
- Soft or metal seats.
- Stem.
- Lever or handwheel.

Added to these components and necessary to complete the construction of the valve and make it functional are the following:

- Stop pin.
- Packing gland.
- Gland nut.
- Gaskets.
- Spring washer.
- Thrust bearing.

This particular valve is designed to a combination of specifications BS 5351 and API 6D for floating ball valves. The ends are threaded NPT to ASME B1.20.1 specifications for socket welds to ASME B16.11 specifications.

It is fire safe according to BS 6755 Part 1 or API 6FA.

Ball Valve—Top Entry, Floating Ball, Soft Seated or Metal Seated, for Cryogenic Service (below −50°F)

The valve in Figure 7.6 is of a top entry design, and because of its smaller size, it has a floating ball with soft seats or the option of metal seats. The ends can be socket welds and supplied welded in pup pieces. The fully assembled valve cannot be socket welded, because the excessive

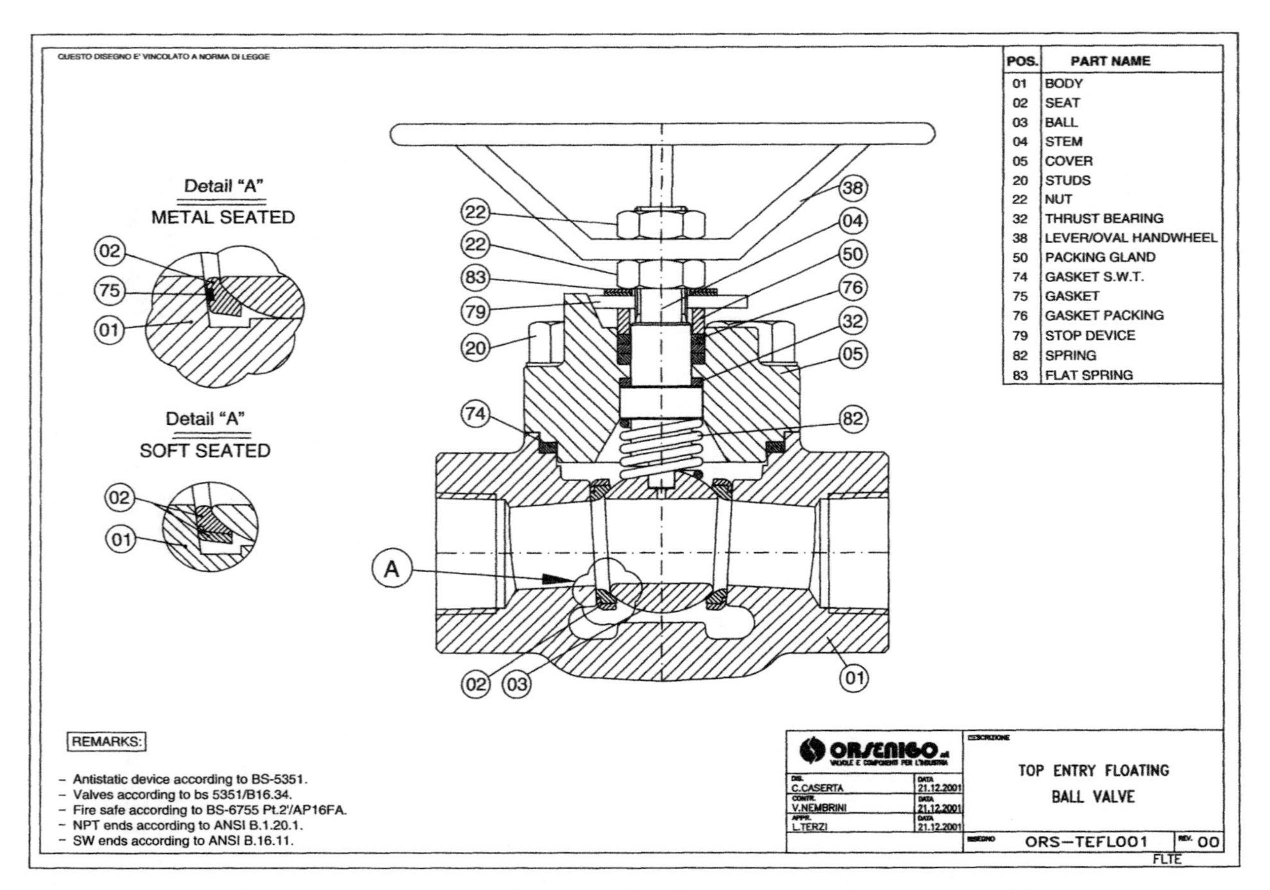

Figure 7.5. Top Entry Floating Ball Valve. (Courtesy of Orsenigo Italy)

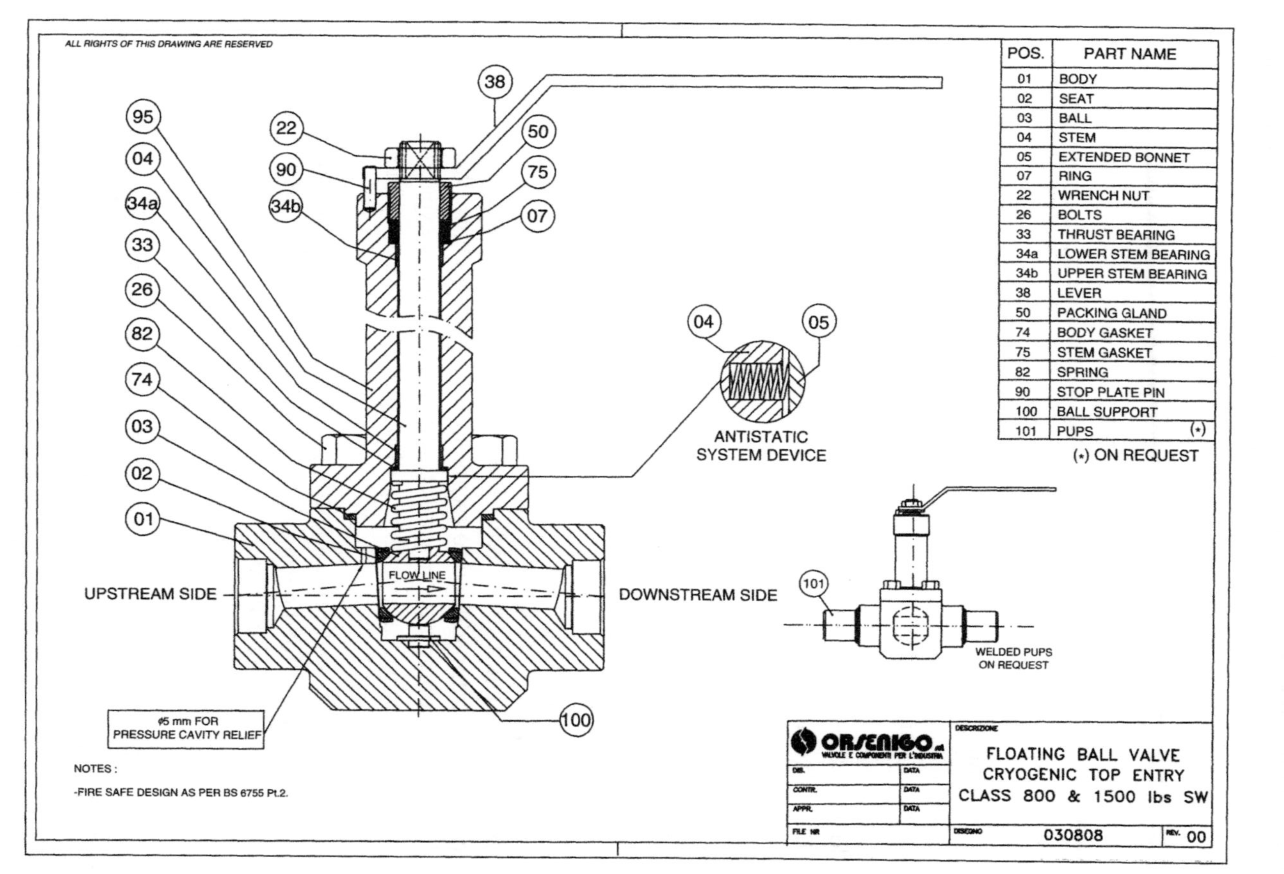

Figure 7.6. Floating Ball Valve, Cryogenic Top Entry, Class 800 and 1500 Socket Weld. (Courtesy of Orsenigo, Italy)

heat required for welding destroys the soft seats. The pup pieces are welded into the valve prior to assembly and the introduction of the soft seats. Also, being top entry it can be repaired in situ.

Components. These are the principal components of the valve:

- Body.
- Studs.
- Ball.
- Soft or metal seats.
- Stem.
- Lever or handwheel.

Added to these components and necessary to complete the construction of the valve and make it functional are the following:

- Stop pin.
- Packing gland.
- Gland nut.
- Gaskets.
- Spring washer.
- Thrust bearing.

This valve also has a cavity relief hole drilled on the upstream side of the valve to allow trapped fluid to escape into the process flow. Without this hole, during decompression, gases trapped in the cavity expand and may squeeze the downstream seats out of position and destroy the complete seal of the valve.

This particular valve is designed to a combination of specifications BS 5351 and API 6D for floating ball valves. The ends are socket welds to ASME B16.11 specifications.

It is fire safe according to specifications BS 6755 or API 6FA.

Ball Valve—Fully Welded, Trunnion Mounted, Soft Seated or Metal Seated

The valve in Figure 7.7 has a fully welded body, which would be specified if the service were so toxic that flanged joints are prohibited or

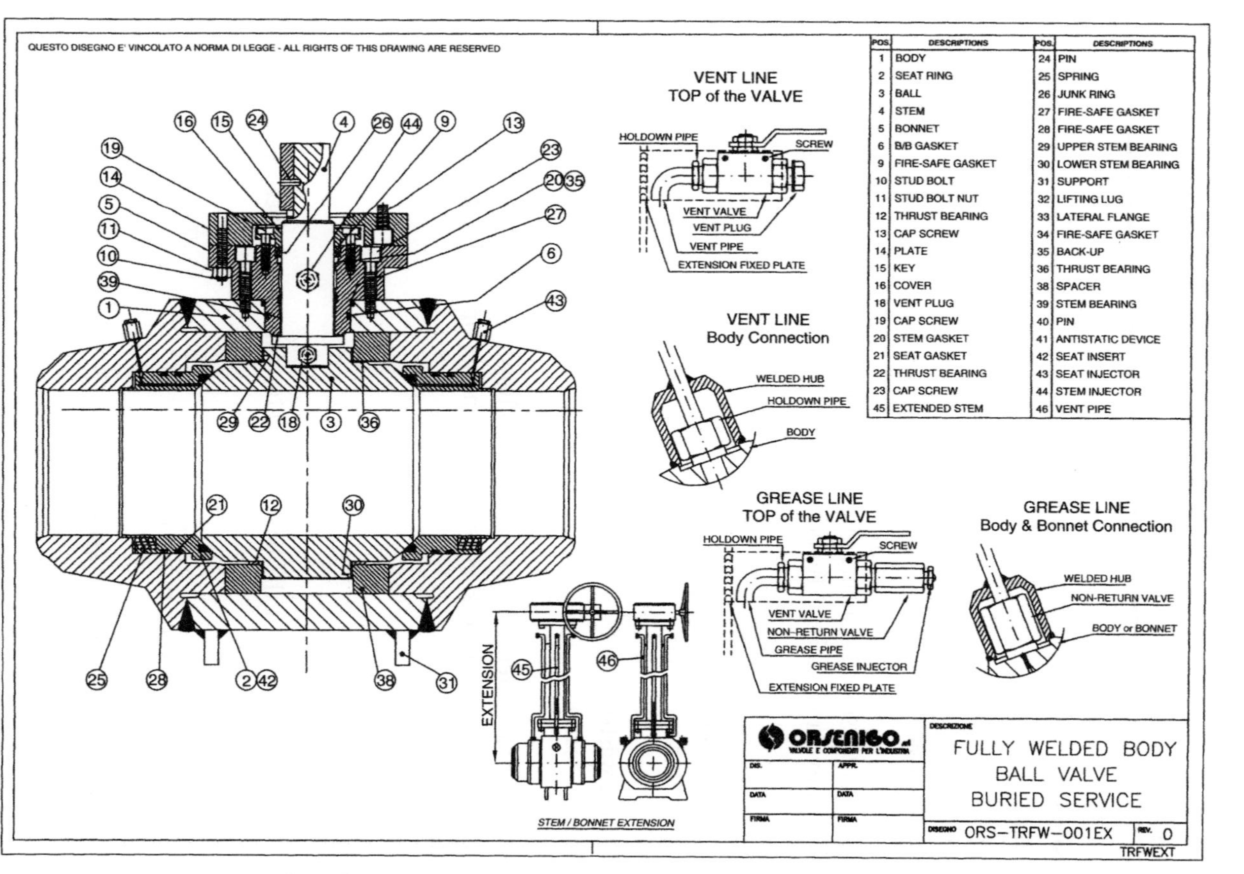

Figure 7.7. Fully Welded Body Ball Valve, Buried Service. (Courtesy of Orsenigo, Italy)

if it were to be buried and not in a pit, because removal for repair is not necessary.

Components. These are the principal components of the valve:

- Body.
- Studs.
- Ball.
- Soft or metal seats.
- Stem.
- Lever or actuator.

Added to these components and necessary to complete the construction of the valve and make it functional are the following:

- Stop pin.
- Packing gland.
- Gland nut.
- Gaskets.
- Spring washer.
- Thrust bearing.

Details show extension stems, which allows operation of the buried valve. The length of the stem varies, depending on the depth to which the valve is buried.

Gate Valves

The manufacturer of the basic gate valve can supply these alternative features in figure 7.8:

- Ends—butt welded or raised faced.
- Port design—full.
- Wedge—solid, flexible, or split.
- Trim—full selection.
- Seats—renewable or seal welded.
- Lantern rings.
- Extended stem for buried or cryogenic service.

- Materials of construction—as per ASME, API, NACE (National Association of Corrosion Engineers), or DIN (Deutsches Institut Fur Normung) specifications.

Special options are also available to suit the client's requirements.

Standards and Specifications. Listed in the figure are all the standards and specifications necessary to construct the valve.

Product Range. Listed in the figure are the various ASME classes and the range that the gate valve can be supplied in by this particular manufacturer. For example, ASME Class 600 is available from 2 to 54 in. with a bolted bonnet and from 2 to 54 in. with a pressure seal.

Globe Valves

This is a basic globe valve, and Figure 7.9 outlines the alternative features that can be supplied, which include the following:

- Ends—butt weld, raised faced, ring-type joint, or socket weld.
- Integral seat.
- Trim—full selection.
- Disc—plug, ball, or needle.
- Extended stem for buried or cryogenic service.
- Materials of construction—as per ASME, API, NACE, or DIN specifications.

Special options are also available to suit the client's requirements.

Standards and Specifications. Listed in the figure are all the standards and specifications necessary to construct the valve.

Product Range. Listed in the figure are the various ASME classes and the range in which the gate valve can be supplied by this manufacturer. For example, ASME Class 600 valves are available from $^3/_8$ to 24 in. with a bolted bonnet and from $^3/_8$ to 24 in. with a pressure seal.

Gate valves

Features

- BW and RF ends.
- Full port design.
- Solid, flexible, or split wedge available.
- Full range of body, bonnet, and trim materials.
- Renewable or seal welded seat rings available
- Anti-friction bearing yoke sleeve, for greater sizes
- Lantern ring and leak off pipe upon request.
- Extended stem (buried valves).
- Criogenic design.
- ASME/API/NACE/DIN materials.
- Special service / heavy duty valves.
- Non-standard valves designed according to customer specifications.

Standards and specifications

- ASME B16.5, Steel pipe flanges and flange fittings.
- ASME B16.10, Face-to-face and end-to-end dimensions of valves.
- ASME B16.25, Buttwelding ends.
- ASME B16.34, Valves - Flanged, threaded, and welding end.
- API 6A, Wellhead and christmas tree equipment.
- API 6D, Specification for pipeline valves.
- MSS-SP-6, Standard finish for contact faces of pipe flanges and connecting-end flanges of valves and fittings.
- MSS-SP-25, Standard marking system for valves, flanges, fittings, and unions.
- MSS-SP-55, Quality standard for steel castings for valves, flanges, and fittings and other piping components.
- NACE MR0175, Sulfide stress cracking resistant metallic materials for oilfield equipment.

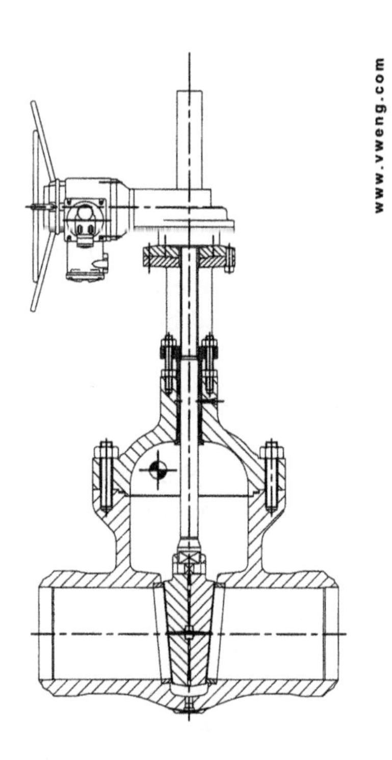

Bolted bonnet & pressure seal

Product range

	Bolted bonnet		Pressure seal	
ASME Class	From	To	From	To
150	2"	72"	—	—
300	2"	54"	—	—
600	2"	54"	2"	54"
900	2"	36"	2"	36"
1500	2"	24"	2"	24"
2500	2"	20"	2"	24"
4500	—	—	2"	24"

VECTOR & WELLHEADS ENGINEERING, S.L.

Figure 7.8. Gate Valve, Bolted Bonnet and Pressure Seal. (Courtesy of Vector and Wellheads Engineering)

Globe valves

www.vweng.com

Features
BW, RF, RTJ, SW, and threaded ends.
Integral seat.
Full range of body, bonnet, and trim materials.
Anti-friction bearing yoke sleeve, 10" and larger.
Plug, ball, or needle disc available.
Extended stem (buried valves).
Criogenic design.
ASME/API/NACE/DIN materials.
Special service / heavy duty valves.
Non-standard valves designed according to customer specifications.

Standards and specifications
ASME B16.5, Steel pipe flanges and flange fittings.
ASME B16.10, Face-to-face and end-to-end dimensions of valves.
ASME B16.25, Buttwelding ends.
ASME B16.34, Valves - Flanged, threaded, and welding end.
API 6A, Wellhead and christmas tree equipment.
API 6D, Specification for pipeline valves.
MSS-SP-6, Standard finish for contact faces of pipe flanges and connecting-end flanges of valves and fittings.
MSS-SP-25, Standard marking system for valves, flanges, fittings, and unions.
MSS-SP-55, Quality standard for steel castings for valves, flanges, and fittings and other piping components.
MSS-SP-84, Steel valves - Socket welding and threaded ends.
NACE MR0175, Sulfide stress cracking resistant metallic materials for oilfield equipment.

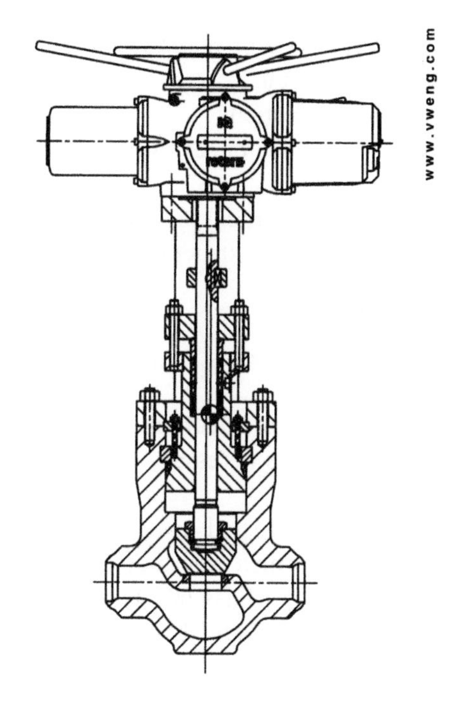

Bolted bonnet & pressure seal

Product range

	Bolted bonnet		Pressure seal	
ASME Class	From	To	From	To
150	3/8"	24"	—	—
300	3/8"	24"	—	—
600	3/8"	24"	2"	24"
900	3/8"	24"	2"	24"
1500	3/8"	24"	2"	24"
2500	3/8"	16"	2"	16"
4500	—	—	2"	6"

VECTOR & WELLHEADS ENGINEERING, S.L.

Figure 7.9. Globe Valve, Bolted Bonnet and Pressure Seal. (Courtesy of Vector and Wellheads Engineering)

Check Valve

This is a basic check valve, and Figure 7.10 outlines the alternative features that can be supplied, which include the following:

- Ends—butt weld or raised faced.
- Port design—full.
- Trim—full selection.
- Seats—renewable or seal welded.
- Antirotation disc.
- Horizontal or vertical placement.
- Materials of construction—as per ASME, API, NACE, or DIN specifications.

Special options are also available to suit the clients requirements.

Standards and Specifications. Listed in the figure are all the standards and specifications necessary to construct the valve.

Product Range. Listed in the figure are the various ASME classes and the ranges in which this manufacturer can supply the gate valve. For example, ASME Class 600 is available from 2 to 54 in. with a bolted bonnet and from 2 to 54 in. with a pressure seal.

Control Valve

Because of its design, the globe pattern is the most suitable valve to control fluids for a wide range of pressures and temperatures and the most commonly specified. The example shown in Figure 7.11 has a butt-weld end, and its design allows it to be maintained without removing it from the line.

Although they are available in sizes above 16 in., for commercial reasons, at the larger sizes a butterfly valve is often specified, for the saving on space and weight.

Check valves

www.vweng.com

Features

BW and RF ends.
Full port design.
Full range of body, bonnet, and trim materials.
Renewable or seal welded seat rings available.
Anti-rotation disc.
Horizontal or vertical service.
Criogenic design.
ASME/API/NACE/DIN materials.
Special service / heavy duty valves.
Non-standard valves designed according to customer specifications.

Standards and specifications

ASME B16.5, Steel pipe flanges and flange fittings.
ASME B16.10, Face-to-face and end-to-end dimensions of valves.
ASME B16.25, Buttwelding ends.
ASME B16.34, Valves - Flanged, threaded, and welding end.
API 6A, Wellhead and christmas tree equipment.
API 6D, Specification for pipeline valves.
MSS-SP-6, Standard finish for contact faces of pipe flanges and connecting-end flanges of valves and fittings.
MSS-SP-25, Standard marking system for valves, flanges, fittings, and unions.
MSS-SP-55, Quality standard for steel castings for valves, flanges, and fittings and other piping components.
NACE MR0175, Sulfide stress cracking resistant metallic materials for oilfield equipment.

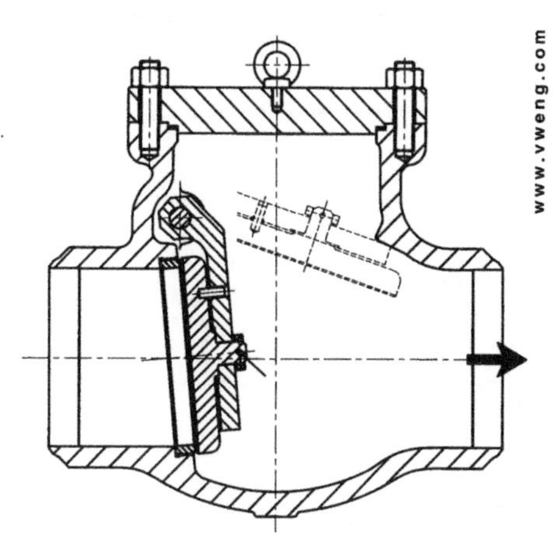

Bolted bonnet & pressure seal

Swing check product range

	Bolted bonnet		Pressure seal	
ASME Class	From	To	From	To
150	2"	64"	—	—
300	2"	54"	—	—
600	2"	54"	2"	54"
900	2"	36"	2"	36"
1500	2"	24"	2"	24"
2500	2"	20"	2"	20"
4500	—	—	2"	8"

Tilting disc product range

	Bolted bonnet		Pressure seal	
ASME Class	From	To	From	To
150	2"	36"	—	—
300	2"	36"	—	—
600	2"	24"	2"	30"
900	2"	20"	2"	24"
1500	2"	18"	2"	24"
2500	2"	12"	2"	16"
4500	—	—	2"	8"

VECTOR & WELLHEADS ENGINEERING, S.L.

Figure 7.10. Check Valve, Bolted Bonnet and Pressure Seal. (Courtesy of Vector and Wellheads Engineering)

Control valves

www.vweng.com

Heavy duty, cage guided

Design and features

Cage guided valves have been designed for sizes up to 26" (DN650) and pressure class ratings, depending on sizes, up to 2500# (PN400). They feature unbalanced plug for sizes up to 4" and balanced plug for sizes 3" and larger. The use of balanced plugs allows, however, Class IV and V seat leak rates, whichever the pressure or the temperature. Also, if temperature does not exceed 250 °C (482 °F), Class VI (bubble tight) can be given as an option.

The use of balanced plugs eliminates the need for oversized actuators, thus reducing weight and cost. Body can be straight or angle type, two or three ways. Bellows sealed or other special bonnets are also included in this versatile range of valves. They have been designed to be operated by pneumatic, electric, or electrohydraulic actuators, including any type of accessories, and accepting any type of control signals. When electric actuators/accessories are used, all kind of protections can be given, including explosion proof or intrinsec safety.

Special designs have been developed, mainly for use in power plants where high pressure and high temperature are usual conditions. They have been used successfully in power plants, gas, oil, fertilizers, and other process industries. Noise and vibration have been reduced, and valve internals life have been extended as well.

Low noise trims and diffusers can be supplied as an integral part of our design. Downstream plates or diffusers can be combined to give further noise reduction, while improving valve performance.

Vector & Wellheads Engineering control valves offer the best combination of design and materials to cope with the most severe operating conditions.

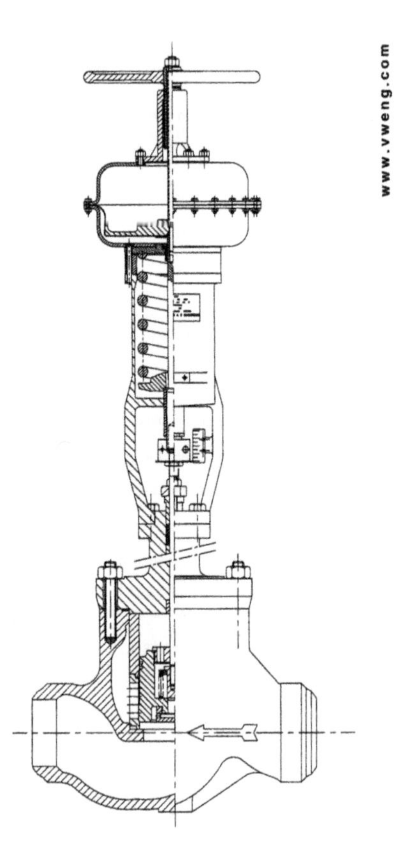

Globe, angle, and three-way product range

Body material	Body size		Pressure class		Ends
	ASME	DIN	ASME	DIN	
Cast steel	3/4" thru 2"	DN 20 thru DN 50	150 thru 2500	PN 10 thru PN 400	SW, BW, FLG
Cast steel	3" thru 26"	DN 80 thru DN 650	150 thru 2500	PN 10 thru PN 400	BW, FLG
Forged steel	3/4" thru 2"	DN 20 thru DN 50	150 thru 2500	PN 10 thru PN 400	SW, BW
Forged steel	3" thru 16"	DN 80 thru DN 400	150 thru 2500	PN 10 thru PN 400	BW

VECTOR & WELLHEADS ENGINEERING, S.L.

Figure 7.11. Control valve. (Courtesy of Vector and Wellheads Engineering)

8

GLOSSARIES AND ABBREVIATIONS

Piping material engineering touches on many topics, and I have included glossaries on several subjects to assist the reader with terminology in several areas when reading this and other technical books:

- Steel.
- Welding.
- Refinery.
- Piping.
- Elastomers and polymers.
- Abbreviations for nonmetals.

1. STEEL GLOSSARY

Piping engineers must have a basic knowledge of the various methods of manufacturing steel products to enable them to specify and evaluate piping components for process plants. It is important, however, to remember that this is a specialist area, and for complex decision making, the additional experience of a metallurgist must be sought.

The following terms are commonly used in the production of steel and the manufacturing of various products, such as pipe, bar, wire, castings, and forgings. It will be useful guide when referring to specific steel material specifications.

Accelerated cooling. The cooling of a plate with water immediately following the final rolling operation. Generally the plate is water cooled from about 1400°F to approximately 1100°F.

Acid brittleness. Brittleness resulting from the pickling of steel in acid.

Activation. The changing of a passive surface of a metal to a chemically active state. In contrast with *passivation.*

Age hardening. Slow, gradual changes that take place in properties of steels after the final treatment. These changes bring about a condition of increased hardness, elastic limit, and tensile strength with, as a consequence, a loss in ductility.

Air cooling. The cooling of the heated metal in the open air.

AISI (American Iron and Steel Institute). A North American trade association with 50 member companies and over 100 associate members.

Alkaline. Having the properties of an alkali, which includes a pH greater than 7.

Alloying element. Any metallic element added during the making of steel for the purpose of increasing corrosion resistance, hardness, or strength. The metals used most commonly as alloying elements in stainless steel include chromium, nickel, and molybdenum.

Alloy steel. An iron-based mixture is considered to be an alloy steel when manganese is greater than 1.65%, silicon over 0.5%, copper above 0.6%, or other minimum quantities of alloying elements such as chromium, nickel, molybdenum, or tungsten are present. An enormous variety of distinct properties can be created for the steel by substituting these elements in the recipe. Addition of such alloying elements is usually to increased hardness, strength, or chemical resistance.

Aluminum (Al). Element no. 13 of the periodic system; atomic weight 26.97; silvery white metal of valence 3; melting point 1220°F; boiling point approximately 4118°F; ductile and malleable; stable against normal atmospheric corrosion but attacked by both acids and alkalis. Aluminum is used extensively in articles requiring lightness, corrosion resistance, electrical, conductivity. Its principal functions is as an alloy in steel making, because it deoxidizes efficiently and restricts grain growth (by forming dispersed oxides or nitrides). It is an alloying element in nitriding steel.

Aluminum killed steel. A steel where aluminum has been used as a deoxidizing agent.

Amalgam. An alloy of mercury with one or more other metals.

Annealing (solution annealing). A process of heating cold stainless steel to obtain maximum softness and ductility, which also produces a homogeneous structure (in austenitic grades) or a 50/50 mixture of austenite and ferrite (in duplex grades). It relieves stresses that built up during cold working and ensures maximum corrosion resistance. Annealing can produce scale on the surface that must be removed by pickling.

Antipitting agent. An addition agent for electroplating solutions to prevent the formation of pits or large pores in the electrodeposit.

Austenitic stainless steel. Nonmagnetic stainless steels that contain nickel and chromium sufficient to develop and retain the austenitic phase at room temperature. Austenitic stainless steels are the most widely used category of stainless steel.

Bark. Surface of metal, under the oxide-scale layer, resulting from heating in an oxidizing environment. In steel, such bark always suffers from decarbonization.

Bars. Stainless steel formed into long shapes from billets. They can be rounds, squares, hexagons, octagons, or flats, either hot or cold finished.

Basic oxygen process. A steel-making process wherein oxygen of the very highest purity is blown onto the surface of a bath of molten iron contained in a basic lined and ladle-shaped vessel. The melting cycle duration is extremely short with quality comparable to the *open hearth process*.

Bath annealing. Immersion in a liquid bath held at an assigned temperature. When a lead bath is used, the process is known as lead annealing.

Bend tests. Tests used to assess the ductility and malleability of stainless steel subjected to bending.

Bessemer process. A process for making steel by blowing air through molten pig iron contained in a refractory lined vessel so that the impurities are thus removed by oxidation.

Beveling. The end preparation for field welding pipe.

Billet. A semi-finished steel form that is used for "long" products: bars, channels, or other structural shapes. A billet is different from a slab because of its outer dimensions; billets are normally 2–7 in. square, while slabs are 30–80 in. wide and 2–10 in. thick. Both shapes are generally continually cast, but they may differ greatly in their chemistry.

Blast furnace. A towering cylinder lined with heat-resistant (refractory) bricks, used by integrated steel mills to smelt iron from its ore. Its name comes from the "blast" of hot air and gases forced up through the iron ore, coke, and limestone that load the furnace.

Blister. A defect in metal produced by gas bubbles, either on the surface or formed beneath the surface while the metal is hot or plastic. Very fine blisters are called "pinhead" or "pepper" blisters.

Bloom. A semi-finished steel form whose rectangular cross-section is more than 8 in. This large cast steel shape is broken down in the mill to produce the familiar I-beams, H-beams, and sheet piling. Blooms are also part of the high-quality bar manufacturing process: Reduction of a bloom to a much smaller cross-section can improve the quality of the metal.

Blooming mill. A hot rolling mill that takes continuously cast slabs or ingots and processes them into blooms.

Blowhole. A cavity produced during the solidification of metal by evolved gas, which in failing to escape, is held in pockets.

Blowpipe. A device for mixing and burning gases to produce a flame for welding, brazing, bronze welding, cutting, heating, and similar operations.

Boron (B). Element no. 5 of the periodic system; atomic weight 10.82. It is gray in color, ignites at about 1112°F, and burns with a brilliant green flame, but its melting point in a nonoxidizing atmosphere is about 4000°F. Boron is used in steel in minute quantities for one purpose only—to increase the ability to harden as in case hardening and to increase strength and hardness penetration.

Brass. Copper base alloy in which zinc is the principal added element. Brass is harder and stronger than either of its alloying elements copper or zinc, is malleable and ductile, develops high tensile strength with cold working, and is not heat treatable for development of hardness.

Brass (cartridge). 70% copper, 30% zinc. This is one of the most widely used of the copper-zinc alloys; it is malleable and ductile, has excellent cold-working but poor hot working and poor machining properties, and develops high tensile strength with cold working.

Brass (yellow). 65% copper, 35% zinc. Known as "high brass" or "two to one brass," it is a copper-zinc alloy yellow in color. Formerly widely used but now largely supplanted by *Brass (cartridge)*.

Brazing. Brazing and soldering are techniques for joining metals in the solid state by means of fusible filler metal with a melting point well below that of the base metal.

Brinell hardness (test). A standard method of measuring the hardness of certain metals. The smooth surface of the metal is subjected to indentation by a hardened steel ball under pressure or load. The diameter of the resultant indentation, in the metal surface, is measured by a special microscope, and the Brinell hardness value read from a chart or calculated formula.

Brinell hardness number (HB). A measure of hardness determined by the Brinell hardness test, in which a hard steel ball under a specific load is forced into the surface of the test material. The number is derived by dividing the applied load by the surface area of the resulting impression.

Brittle fracture. A fracture that has little or no plastic deformation.

Bronze. Primarily an alloy of copper and tin, but the name is now applied to other alloys not containing tin, such as aluminum, bronze, manganese bronze, and beryllium bronze.

Burr. A subtle ridge on the edge of strip stainless steel resulting from cutting operations, such as slitting, trimming, shearing, or blanking. For example, as a stainless steel processor trims the sides of the sheet stainless steel parallel or cuts a sheet of stainless steel into strips, its edges will bend with the direction of the cut.

Butt welding. Joining two specially prepared edges or ends by placing one against the other and welding them.

Calcium (Ca). In the form of calcium silicate, it acts as a deoxidizer and degasifier when added to steel. Recent developments have found that carbon and alloy steels modified with small amounts of calcium show improved machinability and longer tool life. Transverse ductility and toughness are also enhanced.

Capped steel. Semiskilled steel cast in a bottle-top mold and covered with a cap fitting into the neck of the mold. The cap causes the top metal to solidify. Pressure is built up in the sealed-in molten metal and results in a surface condition much like that of rimmed steel.

Carbide. A compound of carbon with one or more metallic elements.

Carbon (C). Element no. 6 of the periodic system; atomic weight 12.01; has three allotropic modifications, all nonmetallic. Carbon is preset in practically all ferrous alloys and has a tremendous effect on the properties of the resultant metal. Carbon is also an essential compound of the cemented carbides. Its metallurgical use, in the form of coke, for reduction of oxides, is extensive.

Carbonitriding. A case-hardening process in which steel components are heated in an atmosphere containing both carbon and nitrogen.

Carbon range. In steel specifications, the carbon range is the difference between the minimum and maximum amount of carbon acceptable.

Carbon steel. A steel containing only residual quantities of elements other than carbon, except those added for deoxidization or to counter the deleterious effects of residual sulfur. Silicon is usually limited to about 0.60% and manganese to about 1.65%. Also termed "plain carbon steel," "ordinary steel," and "straight carbon steel."

Carburization (cementation). Adding carbon to the surface of iron-base alloys by absorption through heating the metal at a temperature below its melting point in contact with carbonaceous solids, liquids, or gasses. The oldest method of case hardening.

Case hardening. Hardening a ferrous alloy to make the outside (case) much harder than the inside (core). This can be done by carburizing, cyaniding, nitriding, carbonitriding, induction hardening, and flame hardening. Their application to stainless steel is limited wherever they decrease corrosion resistance.

Casting. (1) An object at or near finished shape obtained by solidification of a substance in a mold. (2) Pouring molten metal into a mold to produce an object of desired shape.

Cast iron. Iron containing more carbon than the solubility limit in austenite (about 2%).

Cast steel. Steel in the form of castings, usually containing less than 2% carbon.

Cathodic corrosion. Corrosion caused by a reaction of an amphoteric metal with the alkaline products of electrolysis.

Cathodic inhibitor. A chemical substance that prevents or slows a cathodic or reduction reaction.

Cathodic protection. Reducing the corrosion of a metal by making the particular surface a cathode of an electrochemical cell.

Cavitation. The rapid formation and depletion of tiny air bubbles that can damage the material at the solid-liquid interface under conditions of severe turbulent flow.

Cb. Chemical symbol for columbium.

Ce. Chemical symbol for cerium.

Cementite. A compound of iron and carbon, known chemically as iron carbide and having the approximate chemical formula Fe_3C. It is characterized by an orthorhombic crystal structure. When it occurs as a phase in steel, the chemical composition is altered by the presence of manganese and other carbide-forming elements.

Cermet. A powder metallurgy product consisting of ceramic particles bonded with a metal.

Charge. The material loaded into an electric furnace that will melt into a composition that will produce a stainless molten product. Normally recycled scrap, iron, and alloying elements.

Charpy test. A pendulum-type, single-blow impact test in which the specimen, usually notched, is supported at both ends as a simple beam and broken by a falling pendulum. The energy absorbed, as determined by the subsequent rise of the pendulum, is a measure of impact strength or notch toughness.

Chemical analysis. A report of the chemical composition of the elements and their percentages that form a product.

Chemical treatment. An aqueous solution of corrosion-inhibiting chemicals, typically chromate or chromate-phosphate.

Chloride stress corrosion cracking. Cracking due to the combination of tensile stress and corrosion in the presence of water and chlorides.

Chromium (Cr). An alloying element that is the essential stainless steel raw material for conferring corrosion resistance. A film that naturally forms on the surface of stainless steel self-repairs in the presence of oxygen if the steel is damaged mechanically or chemically and thus prevents corrosion from occurring.

Chromium-nickel steel. Steel usually made by the electric furnace process in which chromium and nickel participate as alloying elements. The stainless steel of 18% chromium and 8% nickel are the better known of the chromium-nickel types.

Clad metal. A composite metal containing two or three layers that have been bonded together. The bonding may have been accomplished by corolling, welding, heavy chemical deposition, or heavy electroplating.

Cobalt (Co). Element no. 27 of the periodic system; atomic weight 58.94. A gray magnetic metal of medium hardness, it resists corrosion like nickel, which it resembles closely; melting point is 2696°F, boiling point is about 5250°F, specific gravity is 8.9. It is used as the matrix metal in most cemented carbides and is occasionally electroplated instead of nickel, the sulfate being used as electrolyte. Its principal function is as an alloy in tool steel; it contributes to red hardness by hardening ferrite.

Coefficient of expansion. The ratio of change in length, area, or volume per degree to the corresponding value at a standard temperature.

Coils. A sheet of stainless steel that has been rolled into a coil to facilitate transportation and storage.

Cold-finished steel bars. Hot-rolled carbon steel bars with a higher surface quality and strength produced from secondary cold reduction.

Cold forming (cold working). Any mechanical operation that creates permanent deformation, such as bending, rolling, or drawing, performed at room temperature that increases the hardness and strength of the stainless steel.

Cold-rolled finish. Finish obtained by cold rolling plain pickled sheet or strip with a lubricant resulting in a relatively smooth appearance.

Cold-rolled products. Flat-rolled products for which the required final thickness has been obtained by rolling at room temperature.

Cold-rolled strip (sheet). Sheet steel that has been pickled and run through a cold-reduction mill. Strip has a final product width of approximately 12 in., while sheet may be more than 80 in. wide. Cold-rolled sheet is considerably thinner and stronger than hot-rolled sheet, so it sells for a premium.

Cold rolling. Rolling metal at a temperature below the softening point of the metal to create strain hardening (work hardening). Same as cold reduction, except that the working method is limited to rolling. Cold rolling changes the mechanical properties of strip and produces certain useful combinations of hardness, strength, stiffness, ductility, and other characteristics known as tempers.

Cold treatment. Exposing steel to suitable subzero temperatures (−85°C or −120°F) to obtain desired conditions or properties, such as dimensional or microstructural stability. When the treatment involves the transformation of retained austenite, it is usually followed by tempering.

Columbium (Cb). Element no. 41 of the periodic system; atomic weight 92.91. It is steel gray in color and has a brilliant luster. Specific gravity is 8.57. Melting point is at about 4379°F. It is used mainly in the production of stabilized austenitic chromium-nickel steels, also to reduce the air-hardening characteristics in plain chromium steels of the corrosion resistant type.

Commercial bronze. A copper-zinc alloy (brass) containing 90% copper and 10% zinc, used for screws, wire, hardware, and the like. Although termed "commercial bronze" it contains no tin. It is somewhat stronger than copper and has equal or better ductility.

Commercial-quality steel sheet. Normally, to a ladle analysis of carbon limited at 0.15 maximum. A standard-quality carbon steel sheet. The ladle analysis is taken when the steel is in a molten state.

Consumption. The physical use of stainless steel by end users. Consumption predicts changes in inventories, unlike demand figures.

Continuous casting. Processes of pouring stainless steel into a billet, bloom, or slab directly from the furnace. This process avoids the need for large, expensive mills and also saves time because the slabs solidify in minutes rather than the several hours it takes it for an ingot to form.

Continuous furnace. Furnace, in which the material being heated moves steadily through the furnace.

Continuous pickling. Passing sheet or strip metal continuously through a series of pickling and washing tanks.

Continuous strip mill. A series of synchronized rolling mill stands in which coiled flat rolled metal entering the first pass (or stand) moves in a straight line and is continuously reduced in thickness (not width) at each subsequent pass. The finished strip is recoiled on leaving the final or finishing pass.

Continuous weld. A weld extending along the entire length of a joint.

Controlled atmosphere. A gas or mixture of gases in which steel is heated to produce or maintain a specific surface condition. Controlled atmosphere furnaces are widely used in the heat treatment of steel, as scaling and decarburization of components is minimized by this process.

Controlled-atmosphere furnaces. A furnace used for bright annealing into which specially prepared gases are introduced to maintain a neutral atmosphere so that no oxidizing reaction between metal and atmosphere takes place.

Controlled cooling. A process by which steel is cooled from an elevated temperature in a predetermined manner to avoid hardening, cracking, and internal damage or to produce desired microstructure or mechanical properties.

Cooling stresses. Stresses develop by uneven contraction or external constraint of metal during cooling; also those stresses resulting from localized plastic deformation during cooling and retained.

Copper (Cu). Element no. 29 of the periodic system; atomic weight 63.57. A characteristically reddish metal of bright luster, highly malleable and ductile, and having high electrical and heat conductivity; melting point is 1981°F; boiling point is 4237°F; specific gravity is 8.94. Universally used in the pure state as sheet, tube, rod, and wire, and (see *Brass* and *Bronze*) as an alloy with other metals.

Corrosion. The attack on metals by chemical agents converting them to nonmetallic products. Stainless steel has a passive film created by the presence of chromium (and often other alloying elements, such as nickel and molybdenum) that resists this process.

Corrosion embrittlement. The brittleness caused in certain alloys by exposure to a corrosive environment. Such material is usually susceptible to the intergranular type of corrosion attack.

Corrosion fatigue. Fatigue that arises when alternating or repeated stress combines with corrosion. The severity of the action depends on the range and frequency of the stress, the nature of the corroding condition, and the time under stress.

Corrosion potential. The potential of a corroding surface in an electrolyte relative to a reference electrode under open-circuit conditions.

Corrosion rate. The rate at which an object corrodes.

Corrosion resistance. A metal's ability to resist corrosion in a particular environment.

Crack. A longitudinal discontinuity produced by fracture. Cracks may be longitudinal, transverse, edge, crater, center line, fusion zone underhead, weld metal, or parent metal.

Creep. The flow or plastic deformation of metals held for long periods of time at stresses lower than the normal yield strength. The effect is particularly important if the temperature of stressing is above the recrystallization temperature of the metal.

Creep limit. (1) The maximum stress that causes less than a specified quantity of creep in a given time. (2) The maximum nominal stress under which the creep strain rate decreases continuously with time under a constant load and at a constant temperature. Sometimes used synonymously with *creep strength.*

Creep strength. (1) The constant nominal stress that causes a specified quantity of creep in a given time at constant temperature. (2) The constant nominal stress that causes a specified creep reaction at constant temperature.

Crevice corrosion. Corrosion of a metal surface that is fully shielded from the environment but corrodes because it is so close to the surface of another metal.

Critical cooling rate. The minimum rate of continuous cooling just sufficient to prevent undesired transformations. For steel, the slowest rate at which it can be cooled from above the upper critical temperature to prevent the decomposition of austenite at any temperature above the Ms. Ms is the temperature at which martensitic transformation starts.

Critical range. A temperature range in which an internal change takes place within a metal. Also termed *transformation range.*

Critical surface. Intended for material applied to critical exposed or painted applications where cosmetic surface imperfections are objectionable. The prime side surface is free of repetitive imperfections, gouges, scratches, scale, and slivers. This surface can be furnished only as a pickled product.

Critical temperature. The temperature at which some phase change occurs in a metal during heating or cooling, that is, the temperature at which an arrest or critical point is shown on heating or cooling curves.

Crucible. A ceramic pot or receptacle made of graphite and clay or other refractory materials and used in the melting of metal. The term is sometimes applied to pots made of cast iron, cast steel, or wrought steel.

Cut to length. Cutting flat-rolled stainless steel into the desired length, then normally shipped flat stacked.

Cyanide hardening. A process of introducing carbon and nitrogen into the surface of steel by heating it to a suitable temperature in a molten bath of sodium cyanide or a mixture of sodium and potassium cyanide, diluted with sodium carbonate, and quenching in oil or water. This process is used where a thin case and high hardness are required.

Cyaniding. Surface hardening of an iron-base alloy article or portion of it by heating at a suitable temperature in contact with a cyanide salt, followed by quenching.

Deburring. Removing the subtle ridge from the edge of strip metal that results from a cutting operation, such as slitting, trimming, shearing, or blanking.

Decarburization. Removal of carbon from the outer surface of iron or steel, usually by heating in an oxidizing or reducing atmosphere. Water vapor, oxygen, and carbon dioxide are strong decarburizes. Reheating with adhering scale is also a strongly decarburizing in action.

Degassing process (in steel making). Removing gases from the molten metal by means of a vacuum process in combination with mechanical action.

Deoxidation. A process used during melting and refining of steel to remove or chemically combine oxygen from the molten steel to prevent porosity in the steel when it is solidified.

Descaling. A process that removes from the surface of the stainless steel the oxide scale that develops from hot operations.

Die casting. The principal processes for casting near net shapes of nonferrous metals, such as zinc, aluminum, and zinc-aluminum alloy.

Drawing (drawn). A forming process that presses metal into or through a die (as in cold drawn wire).

Dry film thickness (DFT). The thickness of the dry paint film.

Ductility. A measurement of the malleability of stainless steel in terms of the amount of deformation it withstands before failure.

Duplex. Stainless steel composed of austenitic and ferretic stainless steels that contain high amounts of chromium and nickel. This combination is stronger than both individual stainless steels. Duplex stainless steels are highly resistant to corrosion and cracking.

Eddy-current testing. Nondestructive testing method in which eddy-current flow is induced in the test object. Changes in the flow caused by variations in the object are reflected into a nearby coil or coils for subsequent analysis by suitable instruments and techniques.

Elastic limit. Maximum stress a material stands before permanent deformation.

Electric arc furnace (EAF). A stainless-steel-producing furnace where scrap generally makes up a high percentage of the charge. Heat is supplied from electricity that arcs from the electrodes to the metal bath. These furnaces may operate on ac or dc.

Electric resistance welded (ERW) **pipe.** Pipe made from strips of hot-rolled stainless steel, which are passed through forming rolls and welded.

Electrogalvanized. Zinc plating process whereby the molecules on the positively charged zinc anode attach to the negatively charged sheet steel. The thickness of the zinc coating is readily controlled.

Electroplating. The production of a thin coating of one metal on another by electrodeposition. It is used extensively in industry and is continuing to enlarge its useful functions. Various plated metals and combinations are being used for different purpose; for example:

Purpose	Plating
Decoration and protection against corrosion	Copper, nickel, and chromium
Protection against corrosion	Cadmium or zinc
Protection against wear	Chromium
Buildup of a part or parts under size	Chromium or nickel
Plate for rubber adhesion	Brass
Protection against carburization and for brazing operations	Copper and nickel

Elongation. A measurement of ductility expressed in terms of the stretch having occurred over a given length on a standard tensile specimen at time of fracture, usually based on an original length of 2 in.

Embrittlement. A material's loss of malleability due to chemical treatment or physical change.

Erosion. The continuous depletion of a material due to mechanical interaction with a liquid, a mulitcomponent fluid, or solid particles carried with the fluid.

Erosion corrosion. An accelerated loss of material concerning corrosion and erosion that results from corrosive material interacting with the material.

Extensometer. An apparatus for indicating the deformation of metal while it is subjected to stress.

Extensometer test. The measurement of deformation during stress in the elastic range, permitting determination of elastic properties, such as proportional limit, proof stress, yield strength by the offset method, and so forth. Requires the use of special testing equipment and testing procedures, such as the use of an extensometer or plotting a stress-strain diagram.

Extra-smooth galvanized. An extra-smooth finish is imparted to hot-dip metallic-coated steel sheet by temper rolling after coating to decrease the surface relief that occurs when the molten coating solidifies. The spangle pattern (grain pattern) is made distinctly less visible by the matte finish imparted by the rolling operation. Most extra-smooth sheet is intended for either prepainted or postpainted applications.

Extrusion. A shaped piece of stainless steel produced by forcing the bloom, bar, or rod through a die of the appropriate shape.

Fabricator. An intermediate product producer that purchases materials and processes them specifically for a particular project.

Fatigue. A condition leading to the eventual fracture of a material due to constant or repeated stress that exerts less pressure than the tensile strength of the material.

Feather. The carbon-rich zone, visible in a flame, extending around and beyond the cone when there is an excess of carbonaceous gas.

Ferritic. Magnetic stainless steels that have a low-carbon content and contain chromium as the main alloying element, usually between 13% and 17%. It is the second most widely used stainless steel. Ferretic stainless steels are generally used in automotive trim and exhaust systems, hot water tanks, and interior architectural trim.

Ferroalloy. Metal products such as ferrochrome, ferromanganese, and ferrosilicon commonly used as raw materials to aid various stages in stainless steel making.

Ferrochrome. A common raw material in stainless steel production. This alloy consists of iron and up to 72% chromium.

Ferrous. Any metal that is composed primarily of iron.

Filler metal. A third material that is melted concurrently with the parent metals during fusion or braze welding. It is usually, but not necessarily, of different composition than the parent metals.

Finish. The final condition of the surface after the last phase of production.

Finishing temperature. The temperature at which hot working is completed.

Flame annealing. A process of softening a metal by the application of heat from a high-temperature flame.

Flame cutting. Oxygen cutting in which the appropriate part of the material to be cut is raised to ignition temperature by an oxy-fuel gas flame.

Flame hardening. A hardening process in which the surface is heated by direct flame impingement, then quenched.

Flatness. Flatness is a measure of a cut length sheet's ability to conform to a flat horizontal surface. Maximum deviation from that surface is the degree to which the sheet is "out of flat". Flatness is often expressed quantitatively in either steepness or I-units.

Flat-rolled stainless steel (flat product). Category of stainless steel that includes shapes such as sheet, strip, and plate.

Flux. An iron cleaning agent that consists of limestone and lime. These products react with impurities in the metallic pool and float to the top of the liquid iron.

Foil. Metal with a maximum width of 0.005 in.

Forging. Forming a hot or cold metal into a fixed shape by hammering, upsetting, or pressing.

Forming. A process that brings about a change in the shape of stainless steel by the application of force (e.g., cold forming, hot forming, wire forming).

Fracture test. Nicking and breaking a bar by means of sudden impact, to enable macroscopic study of the fracture.

Free machining. Adding a small amount of some relatively insoluble element (such as sulfur or selenium) to stainless steel to create a minute and widely distributed soft phase that acts as chip breakers during machining.

Fretting. Action that results in surface damage, especially in a corrosive environment, when there is relative motion between solid surfaces in contact under pressure.

Fretting corrosion. Deterioration at the interface of two contacting surfaces under load, which is accelerated by their relative motion.

Full annealing. Heating the metal to about 100°F above the critical temperature range, followed by soaking at this point and slow cooling below the critical temperature.

Fusion penetration (in fusion welding). The depth to which the parent metal has been fused.

Fusion welding. Any welding process in which fusion is employed to complete the weld.

Fusion zone. The part of the parent metals melted into the weld metal.

Galling. Developing a condition on the rubbing surface of one or both mating parts where excessive friction between high spots results in localized welding with substantial spalling and a further roughening of the surface.

Galvanic corrosion. Accelerated corrosion of a metal because of electrical contact with a more noble metal or nonmetallic conductor in a corrosive electrolyte.

Galvanic furnace. A furnace placed over the strip as it exits the zinc bath to produce a fully alloyed iron-zinc coating. The furnace can be gas fired or induction.

Galvanized steel. Steel coated with a thin layer of zinc to provide corrosion resistance in underbody auto parts, garbage cans, storage tanks, or fencing wire. Sheet steel normally must be cold-rolled prior to the galvanizing stage.

Galvanizing. Coating steel with zinc and tin (principally zinc) for rust proofing. Formerly, for galvanizing, cut length steel sheets were passed singly through a bath of the molten metal. Today's galvanizing processing method consists of uncoiling and passing the continuous length of successive coils either through a molten bath of the metal, termed "hot-dipped galvanizing," or by continuously zinc coating the uncoiled sheet electrolytically, termed "electrogalvanizing."

Gauge. A measure of the thickness of stainless steel.

General corrosion. The term used to describe an attack that proceeds in a relatively uniform manner over the entire surface of a metal. Typically, stainless steels do not exhibit general corrosion.

Grain boundary. The individual crystal units constituting the aggregate structure where the crystalline orientation does not change. The grain boundary is where these individual crystal units meet.

Grain flow. Fiberlike lines appearing on polished and etched sections of forgings, caused by orientation of the constituents of the metal in the direction of work during forging.

Grain growth. An increase in the average size of the grains in polycrystalline metal or alloy, usually a result of heating at elevated temperature.

Grain size. The average diameter of grains in the metal under consideration, or alternatively, the number of grains per unit of area. Since increase in grain size is paralleled by lower ductility and impact resistance, the question of general grain size is of great significance. The addition of certain metals affects grain size; for example, vanadium and aluminum tend to give steel a fine grain. The ASTM has set up a grain size standard for steels, and the McQuaid-Ehn Test was developed as a method of measurement.

Grain size number. An arbitrary number calculated from the average number of individual crystals, or grains, that appear on the etched surface of a specimen.

Granular fracture. A type of irregular surface produced when metal fractures, characterized by a rough, grainy appearance as differentiated from a smooth silky, or fibrous, type. It can be subclassified into transgranular and intergranular forms. This type of fracture is frequently called a "crystalline fracture," but the implication that the metal has crystallized is completely misleading.

Graphite. The polymorph of carbon with a hexagonal crystal structure.

Gray cast iron. A cast iron that gives a gray fracture due to the presence of flake graphite. Often called "gray."

Grinding. Removing material from a workpiece with a grinding wheel or abrasive belt.

Grinding cracks. Shallow cracks formed in the surface of relatively hard materials because of excessive grinding heat or the high sensitivity of the material.

Hammer forging. Forging in which the work is deformed by repeated blows. Compare with *forging*.

Hardenability. The ability of a metal, usually steel, to harden in depth as distinguished from the term *hardness*.

Hardness. The degree to which a metal resists cutting, abrasion, penetration, bending, and stretching. The indicated hardness of metals differs somewhat with the specific apparatus measuring hardness. (See

Brinell hardness, Rockwell hardness, Vickers hardness, scleroscope hardness); *tensile strength* also is an indication of hardness.

Hardness test. Hardness testing consists of pressing an indenter into a flat surface under a perfectly controlled load, then measuring the dimension of the resulting indentation. The three methods most commonly used for stainless steel are the Rockwell B, Rockwell C, and Vickers tests. The higher the number, the harder the material.

Heat. The term referring to a batch of refined stainless steel; a charged oxygen or electric furnace full of stainless steel. A heat of stainless steel can be used to cast several slabs, billets, or blooms.

Heat number. The identification that describes the origin of the slab (heat).

Heat-affected zone (HAZ). The part of a metal that is not melted during cutting, brazing, or welding but whose microstructure and physical properties are altered by the process.

Heat treatment. Altering the properties of stainless steel by subjecting it to a series of temperature changes to increase its hardness, strength, or ductility so that it is suitable for additional applications.

High-strength, low-alloy (HSLA). A specific group of steels in which the strength levels are achieved by the addition of moderate amounts of alloying elements. The most common are columbium, vanadium, or titanium.

High-temperature hydrogen attack. A loss of strength and malleability of steel due to high-temperature reactions of absorbed hydrogen with carbides in the steel, resulting in decarburization and internal fissures.

Homogenizing. Holding at a high temperature to eliminate or decrease chemical segregation by diffusion.

Hooke's law. Stress is proportional to strain in the elastic range. The value of the stress at which a material ceases to obey Hooke's law is known as the *elastic limit.*

Hot-dipped steel. Steel run through a molten zinc-coating bath, followed by an air stream "wipe" that controls the thickness of the zinc finish. Done to fix a rust-resistant coating.

Hot forming. Hot forming operations are used widely in the fabrication of stainless steel to take advantage of its lower resistance to shape change. High temperature reduces steel's yield strength, and this results in a marked lowering of the force required to bring about plastic movement or flow from one shape to another. (hot rolling, hot stretching, etc.).

Hot-rolled sheet. Steel sheet that is processed to its final thickness by rolling at high temperatures on a specially designed hot-rolling facility. Also commonly known as "hot rolled unprocessed."

Hot-rolled sheet nontemper rolled. A U.S. Steel term for the product supplied as a coil directly off the hot-strip mill with no additional processing.

Hot-rolled sheet pickled. A U.S. Steel term for a mill edge coil that is pickled, oiled, and temper rolled with coil ends cropped back to meet gauge tolerances.

Hot-rolled sheet pickled nontemper rolled. A U.S. Steel term for a mill edge coil that is pickled and oiled with coil ends cropped back to meet gauge tolerances.

Hot working. Plastic deformation of metal at a temperature sufficiently high enough to not create strain hardening. The lower limit of temperature for this process is the recrystallization temperature

Hydrogen embrittlement. (1) Brittleness of metal, resulting from the occlusion of hydrogen (usually as a by-product of pickling or by codeposition in electroplating). (2) A condition of low ductility resulting from hydrogen absorption and internal pressure developed subsequently. Electrolytic copper exhibits similar results when exposed to reducing atmosphere at elevated temperature.

Hydrogen-induced cracking (HIC). Stepwise internal cracks that connect adjacent hydrogen blisters on different planes in the metal or to the metal surface.

Hydrogen stress cracking. Cracking of a metal resulting from the combination of hydrogen and tensile stress.

Impact energy (impact value). The amount of energy required to fracture a material, usually measured by means of an *Izod* or *Charpy test*. The type of specimen and testing conditions affect the values and therefore should be specified.

Impact test. Impact testing is used to measure the toughness of a material, corresponding to the energy necessary to cause fracture under shock loading. Low toughness is generally associated with brittle shear fracture, and high toughness with ductile plastic tearing.

Impurities. Elements or compounds whose presence in a material is not desired.

Inclusion. A nonmetallic material in a solid metallic material. Slag or other foreign matter entrapped during welding. The defect is usually more irregular in shape than a gas pore.

Induction hardening. A process of hardening a ferrous alloy by heating it above the transformation range by means of electrical induction, then cooling as required. Quench hardening in which the heat is generated by electrical induction.

Ingot. Semi-finished stainless steel that has been poured into molds and solidified. The molds are then removed, and the stainless steel is ready for rolling or forging.

Integrated mills. Facilities that combine all the stainless steel making facilities from melt shop through hot rolling and cold finishing, to produce mill products.

Intergranular corrosion. Preferential corrosion cracking at or along the grain boundaries of a metal.

Intergranular stress corrosion cracking. Stress corrosion cracking in which the cracking occurs along grain boundaries.

Intermittent weld. A series of welds at intervals along a joint.

Internal oxidation. Formation of oxides beneath the surface of a metal.

Investment casting. (1) Casting metal into a mold produced by surrounding (investing) an expendable pattern with a refractory slurry that sets at room temperature, after which the wax, plastic, or frozen mercury pattern is removed through the use of heat. Also called "precision casting" or the "lost-wax process." (2) A casting made by the process.

Iron (Fe). Element no. 26 of the periodic system; atomic weight 55.85. A magnetic silver-white metal of high tensile strength, ductile and malleable. The melting point of pure iron is about 2795°F. Chemically, iron is chiefly base forming. The principal forms of commercial iron are steel, cast iron, and wrought iron.

Iron-based superalloys. These alloys are at the highest end of the range of temperature and strength. Additives such as chrome, nickel, titanium, manganese, molybdenum, vanadium, silicon, and carbon may be used. These super alloys are also referred to as "super chrome stainless steels."

Iron carbide. One of several substitutes for high-quality, low-residual scrap for use in electric furnace steel making. Iron carbide producers use natural gas to reduce iron ore to iron carbide.

Iron ore. A mineral that contains enough iron to be a factor in stainless steel production.

Izod test. A pendulum type single-blow impact test in which the specimen, usually notched, is fixed at one end and broken by a falling pendulum. The energy absorbed, as measured by the subsequent rise of the pendulum, is a measure of impact strength or notch toughness.

Jigsaw steel. Hardened, tempered, and bright polished with round edges. Carbon content 0.85%. Ranges of sizes 0.039 in. to 393 in. in width and 0.016 in. to 0.039 in. in thickness.

Kerf. The void left after metal has been removed by thermal cutting.

Killed steel. Steel deoxidized with a strong deoxidizing agent, such as silicon or aluminum, to reduce the oxygen content to such a level that no reaction occurs between carbon and oxygen during solidification. The term "killed" indicates that the steel has been sufficiently deoxidized to quiet the molten metal when poured into the ingot mold. The general practice is to use aluminum ferrosilicon or manganese as a deoxidizing agent. A properly killed steel is more uniform as to analysis and comparatively free from aging. However, for the same carbon and manganese content, killed steel is harder than rimmed steel. In general, all steels with above 0.25% carbon are killed, also all forging grades, structural steels from 0.15% to 0.25% carbon, and some special steels in the low-carbon range. Most steels below 0.15% carbon are rimmed steel.

Life cycle costing. An accounting method of costing where expenses are allocated over the life of the product. Life cycle costs are often lower for stainless steel than alternatives despite a higher initial outlay, because stainless products generally last longer and require little maintenance.

Light-gauge stainless steel. A very thin sheet of stainless steel that has been either temper rolled or passed through a cold reduction mill.

Line pipe. A pipe extending over long distances that transports oil, natural gas, and other fluids.

Long products. Category of stainless steel that includes rods, bars, and structural products that are described as long rather than flat.

Low-carbon stainless steel. Stainless steel containing less than 0.03% carbon.

Machinability. The relative ease of machining a metal.

Magnetic-particle inspection. A nondestructive method of inspection for determining the existence and extent of possible defects in ferromagnetic materials. Finely divided magnetic particles, applied to the magnetized part, are attracted to and outline the pattern of any magnetic-leakage fields created by discontinuities.

Malleability. The property that determines the ease of deforming a metal when the metal is subjected to rolling or hammering. The more malleable metals can be hammered or rolled into thin sheet more easily than others.

Magnesium (Mg). Element no. 12 of the periodic system; atomic weight 24.305. Specific gravity 1.77 with a melting point of approximately 1160°F. A silver-white, light, malleable, ductile metallic element that occurs abundantly in nature. The metal is used in metallurgical and chemical processes, in photography, in signaling, and in the manufacture of pyrotechnics because of the intense white light it produces on burning.

Manganese (Mn). Element no. 25 of the periodic system; atomic weight 54.93. Lustrous, reddish-white metal of a hard brittle and, therefore, nonmalleable character. The metal is used in large quantities in the form of Spiegel and ferromanganese for steel manufacture as well as in manganese and many copper-base alloys. Its principal function is as an alloy in steel making. (1) It is a ferrite-strengthening and carbide-forming element. It increases hardenability inexpensively, with a tendency toward embrittlement when too high carbon and too high manganese accompany each other. (2) It counteracts brittleness from sulfur.

Martensitic. Small category of stainless steel characterized by the use of heat treatment for hardening and strengthening. Martensitic stainless steels are plain chromium steels with no significant nickel content. They are utilized in equipment for the chemical and oil industries and in surgical instruments. The most popular martensitic stainless steel is type 410 (a grade appropriate for nonsevere corrosion environments requiring high strength).

Martensitic stainless steel. Has a body centered tetragonal (BCT) structure. These alloys are chromium stainless steels with medium to high carbon levels. They harden slowly in the annealed (soft) condition but can be heat treated to very high tensile strengths.

Matrix. The principal phase or aggregate in which another constituent is embedded.

Matt or matte finish (steel). Not as smooth as a normal mill finish. Produce by etched or mechanically roughened finishing rolls.

Mechanical polishing. A method of producing a specularly reflecting surface by use of abrasives.

Mechanical properties. Those properties of a material that reveal the elastic and inelastic reaction when force is applied or that involve the relationship between stress and strain; for example, the modulus of elasticity, tensile strength, and fatigue limit. These properties have often been designated physical properties, but the term "mechanical properties" is much to be preferred. The mechanical properties of steel depend on its microstructure.

Mechanical working. Plastic deformation or other physical change to which metal is subjected, by rolling, hammering, drawing, or the like to change its shape, properties, or structure.

Medium-carbon steel. Contains from 0.30–0.60% carbon and less than 1.00% manganese. May be made by any of the standard processes.

Melting point. The temperature at which a substance changes form solid to liquid; the temperature at which the liquid and the solid are in equilibrium.

Metal. An opaque, lustrous, elemental substance that is a good conductor of heat and electricity and, when polished, a good reflector of light. Most metals are malleable and ductile and are, in general, denser than other substances.

Metal spraying. A process for applying a coating of metal to an object. The metal, usually in the form of wire, is melted by an oxyhydrogen or oxyacetylene blast or by an electric arc and projected at high speed by gas pressure against the object being coated.

Microstructure. The structure of a prepared surface of a metal as revealed by a microscope at a magnification greater than 10 diameters.

MIG (metal inert gas) **welding.** Inert-gas welding using a consumable electrode (inert-gas metal-arc welding).

Mild steel. Carbon steel containing a maximum of about 0.25% Carbon.

Mill finish. A surface finish produced on sheet and plate; characteristic of the ground finish used on the rolls in fabrication.

Modulus of elasticity (Young's modulus). A measure of the rigidity of metal. The ratio of stress, within proportional limits, to a corresponding strain. Specifically, the modulus obtained in tension or compression is Young's modulus, stretch modulus, or modulus of extensibility; the modulus obtained in torsion or shear is modulus of rigidity, shear modulus, or modulus of torsion; the modulus covering the ratio of the mean normal stress to the change in volume per unit of volume is the bulk modulus. The tangent modulus and secant modulus are not restricted within proportional limits: The former is the slope of the stress-strain curve at a specified point; the latter is the slope of a line from the origin to a specified point on the stress-strain curve. Also called "elastic modulus" and "coefficient of elasticity."

Molybdenum (Mo). Element no. 42 of the periodic system; atomic weight 95.95. Hard, tough metal of grayish white color, becoming very ductile and malleable when properly treated at high temperatures; melting point is 4748°F; boiling point is about 6600°F; specific gravity is 10.2. Pure molybdenum can best be obtained as a black powder, by reduction of molybdenum trioxide or ammonium molybdate with hydrogen. From this powder, ductile sheet and wire are made by powder metallurgy techniques; these are used in radio and related work. Its principal function is as an alloy in steel making: (1) Raises grain-coarsening temperature of austenite. (2) Deepens hardening. (3) Counteracts tendency toward temper brittleness. (4) Raises hot and creep strength, red hardness. (5) Enhances corrosion resistance in stainless steel. (6) Forms abrasion-resisting particles.

NDT. Nondestructive testing.

Nickel (Ni). Element no. 28 of the periodic system; atomic weight 58.69. Silvery white, slightly magnetic metal, of medium hardness and high degree of ductility and malleability and resistance to chemical and atmospheric corrosion; melting point is 2651°F; boiling point is about 5250°F, specific gravity is 8.90. An alloying element used as a raw material for certain classes of stainless steel. Nickel provides high degrees of ductility (ability to change shape without fracture) as well as resistance to corrosion. Approximately 65% of all nickel is used in making stainless steel. Also used for electroplating. Used as an alloying agent, it is of great importance in iron-base alloys in stainless steels and in copper-base alloys such as cupronickel, as well as in nickel-base alloys such as Monel metal. Its principal function is as an alloy in steel making: (1) Strengthens unquenched or annealed steels. (2) Toughens pearlitic-ferritic steels (especially at low temperature). (3) Renders high-chromium iron alloys austenitic.

Nickel-based superalloys. Alloy metal produced for high-performance, high-temperature applications such as nickel-iron-chrome alloys and nickel-chrome-iron alloys.

Nickel steel. Steel containing nickel as an alloying element. Varying amounts are added to increase the strength in the normalized condition to enable hardening to be performed in oil or air instead of water.

Niobium (Nb). Element no. 41 of the periodic system. Also known as Columbium (Cb).

Nitriding. Introducing nitrogen into a solid ferrous alloy by holding at a suitable temperature (below Ac1 for ferritic steels) in contact with a nitrogenous material, usually ammonia of molten cyanide of appropriate composition. Quenching is not required to produce a hard case. Process of surface hardening certain types of steel by heating in ammonia gas at about 935–1000°F, the increase in hardness resulting from surface nitride formation. Certain alloying constituents, principal among them aluminum, greatly facilitate the hardening reaction. In general, the depth of the case is less than with carburizing.

Nitriding steel. Steel particularly suited for the nitriding process; that is, it will form a very hard, adherent surface on proper nitriding (heating in a partially dissociated atmosphere of ammonia gas). Composition is usually 0.20–0.40% carbon, 0.90–1.50% chromium, 0.15–1.00% molybdenum, and 0.85–1.20% aluminum.

Nonferrous metal. Metal or alloy that contains no iron.

Normalizing. A heat treatment applied to steel that involves heating above the critical range followed by cooling in still air. It is performed to refine the crystal structure and eliminate internal stress.

Notch brittleness. A measure of the susceptibility of a material to brittle fracture at locations of stress concentration. For example, in a notch tensile test, a material is said to be notch brittle if its notch strength is less than its tensile strength; otherwise, it is said to be notch ductile.

Notch (impact) **toughness.** An indication of a steel's capacity to absorb energy when a stress concentrator or notch is present. Examples of measurements are Charpy V-notch, dynamic tear, drop-weight, and drop-weight tear tests.

Oil country tubular goods (OCTG). Category of pipe products used by petroleum exploration customers. Labels bearing OCTG are applied to casting, drill pipes, oil well tubing, and the like.

Oil hardening. A process of hardening a ferrous alloy of suitable composition by heating it within or above the transformation range and quenching in oil.

Open-hearth process. Process of making steel by heating the metal in the hearth of a regenerative furnace. In the basic open-hearth steel process, the lining of the hearth is basic, usually magnetite; whereas in the acid open-hearth steel process, an acid material, silica, is used as the furnace lining and pig iron, extremely low in phosphorous (less than 0.04%), is the raw material charged in.

Orange peel (effect). A surface roughening (defect) encountered in forming products from metal stock that has a coarse grain size. It is due to uneven flow or the appearance of the overly large grains, usually the result of annealing at too high a temperature. Also referred to as "pebbles" and "alligator skin."

Ore. An iron-containing material used primarily in the melting furnace.

Oscillating. A method of winding a narrow strip of stainless steel over a much wider roll. This allows for more stainless steel per roll and allows the customer to have longer processing runs.

Oxidation. The addition of oxygen to a compound. Exposure to atmosphere sometimes results in oxidation of the exposed surface, hence a staining or discoloration. This effect is increased with temperature increases. A reaction in which there is an increase in valence resulting from a loss of electrons.

Oxide. Compound of oxygen with another element.

Oxygen-arc cutting. Thermal cutting in which the ignition temperature is produced by an electric arc and cutting oxygen is conveyed through the center of an electrode, which is consumed in the process.

Passivation. When exposed in air, stainless steels passivate (become inactive or less reactive) naturally, due to the presence of chromium. But the time required can vary. To ensure that the passive layer reforms rapidly after pickling, a passivation treatment is performed using a solution of nitric acid and water.

Passive. A characteristic condition of stainless steels which impedes normal corrosion tendencies to the point where the metal remains virtually unattacked, hence passive, to its environment.

Pearlite. A eutectoid transformation product of ferrite and cementite that ideally has a lamellar structure but that is always degenerate to some extent. Lamellar structure resembles mother of pearl. A compound of iron and carbon occurring in steel as a result of the transformation of austenite into aggregations of ferrite and iron carbide.

Peening. Mechanical working of metal by hammer blows or shot impingement.

Phosphorus (P). Element no. 15 of the periodic system; atomic weight 30.98. It is a nonmetallic element occurring in at least three allotropic forms; melting point is 111°F; boiling point is 536°F; specific gravity is 1.82. In steels, it is usually undesirable, with limits set in most specifications. However, it is specified as an alloy in steel to prevent the sticking of light-gauge sheets; to a degree, it strengthens low-carbon steel, increases resistance to corrosion, and improves machinability in free-cutting steels. In the manufacture of phosphor bronze, it is used as a deoxidizing agent.

Physical properties. Those properties familiarly discussed in physics, exclusive of those described under mechanical properties; for example, density, electrical conductivity, coefficient of thermal expansion. This term often has been used to describe mechanical properties, but this usage is not recommended.

Pickling. A process that removes surface scale and oxidation products by immersion in a chemically active solution, such as sulfuric or hydrochloric acid.

Pickling paste. A commercially available product that performs the pickling function when used on the surface of stainless steel.

Pig iron. The name for the melted iron produced in a blast furnace, containing a large quantity of carbon (above 1.5%). Named long ago, when molten iron was poured through a trench in the ground to flow into shallow earthen holes, the arrangement looked like newborn pigs suckling. The central channel became known as the "sow," and the molds were "pigs."

Pipe. Technically, a tube used to transport fluids or gases. However, "pipe" and "tube" are often used interchangeably in steel lexicon, with a given label applied primarily as a matter of historic use.

Pitting. Localized corrosion (in the form of pits) on a metal surface confined to a small area.

Plasticity. The ability of a metal to be deformed extensively without rupture.

Plate. Sheet steel with a width of more than 8 in., with a thickness ranging from 1/4 in. to more than 1 ft.

Plate martensite. Martensite formed, partly in steels containing more than about 0.5% Carbon and solely in steels containing more than about 1.0% Carbon, as lenticular-shape plates on irrational habit planes that are near (225)A, or (259)A in very-high-carbon steels.

Plating. A thin coating of metal laid on another metal.

Plug weld. A weld made by filling a hole in one component of a workpiece to join it to the surface of an overlapping component exposed through the hole.

Polished surface. The finish obtained by buffing with rouge or similar fine abrasive, resulting in a high gloss or polish.

Porosity. The presence of gas pores.

Postheating. Heating weldments immediately after welding, for tempering, stress relief, or control of cooling to prevent formation of a hard or brittle structure.

Postweld heat treatment (PWHT). Also referred to as "stress relief," this process is used to soften the heat-affected zones and relieve residual stresses created during welding.

Powder metals. Fabricating technique in which fine metallic powder is compacted and heated under high pressure to solidify the material.

Precipitation hardening (PH). A small category of stainless steels resembling martenistic stainless steels that have great strength and hardness due to heat treatment.

Protective coating. A temporary adhesive protective film attached to the surface that protects the surface during forming and handling operations and is stripped before final use.

Quench hardening (steel). A process of hardening a ferrous alloy of suitable composition by heating within or above the transformation range and cooling at a rate sufficient to increase the hardness substantially. The process usually involves the formation of martensite.

Quenching. In the heat treatment of metals, the step of cooling metals rapidly to obtain desired properties; most commonly accomplished by immersing the metal in oil or water. In the case of most copper-base alloys, quenching has no effect other than to hasten cooling.

Radiography. A nondestructive method of internal examination in which metal objects are exposed to a beam of X-ray or gamma radiation. Differences in thickness, density, or absorption, caused by internal defects or inclusions, are apparent in the shadow image either on a fluorescent screen or photographic film placed behind the object.

Red brass. A copper-zinc alloy, containing approximately 85% copper and 15% zinc, used for plumbing pipe, hardware, condenser tubes. Because of its color, it is used for vanity cases, coins, plaques, badges, and the like. It is somewhat stronger than commercial bronze and hardened more rapidly by cold working.

Reducing agent. Either natural gas or coal can be used to remove the oxygen from iron ore to produce a scrap substitute. In gas-based processes, the iron ore is heated in a vessel as reformed natural gas passes through. In coal-based processes, iron ore is combined with gasified or ground coal and heated. The oxygen in the ore combines with carbon and hydrogen in the gas or coal, producing reduced, or metallic, iron.

Refining temperature. A temperature, usually just higher than the transformation range, employed in the heat treatment of steel to refine the structure, in particular, the grain size.

Refractory. A heat-resistant material, usually nonmetallic, used for furnace linings and such.

Refractory alloy. A term applied to those alloys that, due to hardness or abrasiveness, present relative difficulty in maintaining close dimensional tolerances.

Refractory brick. Heat-resistant brick. Because its melting point is well above the operating temperatures of the process, refractory bricks line most steel-making vessels that come in contact with molten metal, like the walls of the blast furnace, and sides of the ladles.

Reinforcing bar (rebar). A commodity-grade stainless steel used to reinforce concrete in highway and building structures.

Residual elements. Small quantities of elements unintentionally present in an alloy.

Residuals. The impurities remaining in mini-mill stainless steels resulting from the wide variety of metals entering the process.

Residual stress. Macroscopic stresses that are set up within a metal as the result of nonuniform plastic deformation. This deformation may be caused by cold working or drastic gradients of temperature from quenching or welding. Stress remaining in a metal part or structure as a result of welding.

Resistance welding. A type of welding process in which the workpieces are heated by the passage of an electric current through the contact. Such processes include spot welding, seam or line welding, and percussion welding. Flash and butt welding are sometimes considered resistance welding processes.

Reverse bend test. A bend test in which the side other than that specified for a face bend test is in tension.

Reversing mill. A stand of rolls that passes stainless steel back and forth between the rolls to reduce the stainless steel sheet or plate. The distance between the rolls is reduced after each pass.

Rimmed steel. Low-carbon steel containing sufficient iron oxide to produce continuous evolution of carbon monoxide during ingot solidification, resulting in a case, or rim, of metal virtually free of voids. The rim is of somewhat purer composition than the original metal poured. If the rimming action is stopped shortly after pouring of the ingot is completed, the metal is known as "capped steel." Most steels below 0.15% carbon are rimmed steels. For the same carbon and manganese content, rimmed steel is softer than killed steel.

Rockwell hardness (test). A standard method for measuring the hardness of metals. The hardness is expressed as a number related to the depth of residual penetration of a steel ball or diamond cone (brale) after a minor load of 10 kg has been applied to hold the penetrator in position. This residual penetration is automatically registered on a dial when the major load is removed from the penetrator. Various dial readings combined with different major loads, five scales designated by letters varying from A to H; the B and C scales are most commonly in use.

Rod. Round, thin semi-finished steel length that is rolled from a billet and coiled for further processing. Rod is commonly drawn into wire products or used to make bolts and nails. Rod trains (rolling facilities) can run as fast as 20,000 ft per minute, more than 200 mi an hour.

Roll forming. An operation used in forming sheet. Strips of sheet are passed between rolls of definite settings that bend the sheet progressively into structural members of various contours, sometimes called "molded sections."

Rolling. Reducing the cross-sectional area of metal stock or otherwise shaping metal products through the use of rotating rolls.

Rolling mills. Equipment used for rolling down metal to a smaller size or a given shape, employing sets of rolls the contours of which determine or fashion the product into numerous intermediate and final shapes, such as blooms, slabs, rails, bars, rods, sections, plates, sheets, and strip.

Root (of weld). The zone on the side of the first run farthest from the welder.

Scale (scale removal). The oxide that forms on the surface of stainless steel, after exposure to high temperature.

Scrap. Iron-containing stainless steel material that is normally remelted and recasted into new stainless steel. Home scrap is leftover stainless steel generated from edge trimming and rejects within the mill, also industrial scrap trimmed by stampers and auctioned to buyers.

Seal weld. A weld, not being a strength weld, used to make a seal.

Seamless pipe. Pipe produced from a solid billet that is heated and rotated under pressure. This rotating pressure creates a hole in the middle of the billet, which is then formed into a pipe by a mandrel.

Semi-finished stainless steel. Stainless steel products, such as blooms, billets, or slabs, that are then rolled and processed into beams, bars, sheets, and so forth.

Shearing. Trimming the edges of sheet strip to make them parallel. This is done at either the stainless steel mill or the stainless steel processor.

Sheet. A stainless steel flat rolled product that is under $^{3}/_{16}$ in. thickness and 24 in. and over in width.

Shot blasting. Blast cleaning using stainless steel shot as the abrasive. Not recommended for stainless steel; glass beads should be used.

Shot peening. Stressing the surface layer of a material by bombarding it with a selected medium (usually round steel shot) under controlled conditions.

Sigma phase. An extremely brittle Fe-Cr phase that can form at elevated temperatures in austenitic and ferritic stainless steels.

Silicon (Si). Element no. 14 of the periodic system; atomic weight 28.06. Extremely common element, the major component of all rocks and sands; its chemical reactions, however, are those of a metalloid. Used in metallurgy as a deoxidizing scavenger. Silicon is present, to some extent, in all steels, and deliberately added to the extent of approximately 4% for electric sheets, extensively used in alternating current magnetic circuits. Silicon cannot be electrodeposited.

Skelp. Steel that is the entry material to a pipe mill. It resembles hot-rolled strip, but its properties allow for the severe forming and welding operations required for pipe production.

Skin. A thin surface layer that is different from the main mass of a metal object in composition, structure, or other characteristics.

Slab. A very common type of semi-finished stainless steel usually measuring 6–10 in. thick by 30–85 in. wide and averaging 20 ft long. After casting, slabs are sent to a strip mill where they are rolled and coiled into sheet and plate products.

Slag. The impurities in a molten pool of iron. Flux may be added to congregate the impurities into a slag. Slag is lighter than iron and floats, allowing it to be skimmed off.

Smelter. A processor of mine feed or scrap material (secondary smelter) that produces crude metal.

Solid solution. A solid crystalline phase containing two or more chemical species in concentrations that may vary between limits imposed by phase equilibrium.

Solution heat treatment. Heating a metal to a high temperature and maintaining the temperature long enough for one or more constituents to enter the solid solution. The solution is then cooled rapidly to retain the constitutes within.

Solvent cleaning. The removal of contaminants such as oil, grease, dirt, and salts by cleaning with a solvent, steam, vapor, alkali, or emulsion.

Specialty alloys. Metals with distinct chemical and physical properties. These alloys are produced for very specific applications, considered to be on the low end of superalloys.

Spectograph. An optical instrument for determining the presence or concentration of minor metallic constituents in a material by indicating the presence and intensity of specific wave lengths of radiation when the material is thermally or electrically excited.

Spherodized structure. A microstructure consisting of a matrix containing spheroidal particles of another constituent.

Stainless steel. The term for grades of steel that contain more than 10% chromium, with or without other alloying elements. Stainless steel resists corrosion, maintains its strength at high temperatures, and is easily maintained. For these reasons, it is used widely in items such as automotive and food processing products, as well as medical and health equipment. The most common grades of stainless steel are as follows:

Type 304. The most commonly specified austenitic (chromium-nickel stainless class) stainless steel, accounting for more than half of the stainless steel produced in the world. This grade withstands ordinary corrosion in architecture, is durable in typical food processing environments, and resists most chemicals. Type 304 is available in virtually all product forms and finishes.

Type 316. Austenitic (chromium-nickel stainless class) stainless steel containing 2–3% molybdenum (whereas 304 has none). The inclusion of molybdenum gives 316 greater resistance to various forms of deterioration.

Type 409. Ferritic (plain chromium stainless category) stainless steel suitable for high temperatures. This grade has the lowest chromium content of all stainless steels and therefore is the least expensive.

Type 410. The most widely used martensitic (plain chromium stainless class with exceptional strength) stainless steel, featuring the high level of strength conferred by the martensite. It is a low-cost, heat-treatable grade suitable for nonsevere corrosion applications.

Type 430. The most widely used ferritic (plain chromium stainless category) stainless steel, offering general-purpose corrosion resistance, often in decorative applications.

Steel. An iron-base alloy, malleable in some temperature ranges as initially cast, containing manganese, usually carbon, and often other alloying elements. In carbon steel and low-alloy steel, the maximum carbon is about 2.0%; in high-alloy steel, about 2.5%. The dividing line between low-alloy and high-alloy steels is generally regarded as being at about 5% metallic alloying elements. Steel is differentiated from two general classes of irons: the cast irons, on the high-carbon side, and the relatively pure irons, such as ingot iron, carbonyl iron, and electrolytic iron, on the low-carbon side. In some steels containing extremely low carbon, the manganese content is the principal differentiating factor. Steel usually contains at least 0.25% manganese; ingot iron contains considerably less.

Strain. The amount of elongation, force, or compression that occurs in a metal at a given level of stress. Generally stated in terms of inches elongation per inch of material.

Strength. Properties related to the ability of steel to oppose applied forces. Forms of strength include withstanding imposed loads without a permanent change in shape or structure and resistance to stretching.

Stress. Deforming force to which a body is subjected or the resistance the body offers to deformation by the force.

Stress-corrosion cracking (SCC). Failure by cracking under the combined action of corrosion and stress, either external (applied) or internal (residual). Cracking may be either intergranular or transgranular, depending on the metal and the corrosive medium.

Stress cracking. Occurs during the thermal cutting of high carbon and alloy steels at the cut edges. Proper processing, which may include preheating, prevents this problem.

Stress relief. Low-temperature annealing to remove internal stresses, such as those resulting on a metal from work hardening or quenching.

Stress relieving. Heating to a suitable temperature, holding the temperature long enough to reduce residual stresses, then cooling slowly enough to minimize the development of new residual stresses.

Stress-rupture test. A tension test performed at constant temperature, the load being held at such a level as to cause rupture. Also known as "creep-rupture test."

Strip. A stainless steel flat rolled product that is less than 3/16 in. in thickness and is under 24 in. in width.

Structurals. An architectural stainless steel product group that includes I-beams, H-beams, wide-flange beams, and sheet piling. These products are used in multistory buildings, bridges, vertical highway supports, and so on.

Submerged-arc welding. Metal-arc welding using a bare wire electrode or electrodes; the arc or arcs are enveloped in a flux, some of which fuses to form a removable covering of slag on the weld.

Substrate. The layer of metal underlying a coating, regardless of whether the layer is base metal. Raw material used as an input for steel processing; for example, hot-rolled steel is the substrate for cold-rolling operations.

Sulfide stress cracking. Cracking of a metal under the combined action of tensile stress and corrosion in the presence of water and hydrogen sulphide (a form of hydrogen stress cracking).

Sulfur (S). Element no. 16 of the periodic system; atomic weight 32.06. Nonmetal occurring in a number of allotropic modifications, the most common being a pale-yellow brittle solid. In steel, most commonly encountered as an undesired contaminant. However, it is frequently deliberately added to cutting stock to increase machinability.

Superalloys. Lightweight metal alloys designed specifically to withstand extreme conditions. Conventional alloys are iron based, cobalt based, nickel based, and titanium based.

Superficial Rockwell hardness test. Form of Rockwell hardness test using relatively light loads, which produce minimum penetration. Used for determining surface hardness or hardness of thin sections or small parts or where large hardness impression might be harmful.

Surface-fusion welding. Gas welding in which a carburizing flame is used to melt the surface of the parent metal, which then unites with the metal from a suitable filler rod.

Tantalum (Ta). A by-product of tin processing, this refractory metal is used as a barrier to corrosion of chemical processing and carbide cutting tools and, increasingly, as electronic capacitors and filaments. Melts at 2415°F.

Tapping. Transferring molten metal from a melting furnace to a ladle.

Tarnish. Surface discoloration on a metal, usually from a thin film of oxide or sulfide.

Teeming. Pouring molten metal from a ladle into ingot molds. The term applies particularly to the specific operation of pouring either iron or steel into ingot molds.

Tempering. A process of reheating quench-hardened or normalized steel to a temperature below the transformation range then cooling at any rate desired. The primary purpose of tempering is to impart a degree of plasticity or toughness to the steel to alleviate the brittleness of its martensite.

Tensile strength (test). Also called *ultimate strength,* it is the breaking strength of a material when subjected to a tensile (stretching) force. It is usually measured by placing a standard test piece in the jaws of a tensile machine, gradually separating the jaws, and measuring the stretching force necessary to break the test piece. The tensile strength is commonly expressed as pounds (or tons) per square inch of original cross section.

Test piece. Components welded together in accordance with a specified welding procedure or a portion of a welded joint detached from a structure for testing.

Thermal analysis. A method of studying transformations in metal by measuring the temperatures at which thermal arrests occur.

Thermal cutting. The parting or shaping of materials by the application of heat, with or without a stream of cutting oxygen.

Thermal treatment. Any operation involving the heating and cooling of a metal or alloy in a solid state to obtain the desired microstructure or mechanical properties.

Thermocouple. A device for measuring temperatures by the use of two dissimilar metals in contact; the junction of these metals gives rise to a measurable electrical potential with changes in temperature.

Thermo-mechanical-controlled-processing (TMCP). A term referring to special rolling practices that use controlled-rolling, accelerated cooling, or both.

TIG (tungsten inert gas) **welding.** Inert-gas welding using a nonconsumable electrode (inert-gas tungsten-arc welding).

Tin (Sn). Element no. 50 of the periodic system; atomic weight 118.70. Soft silvery white metal of high malleability and ductility but low tensile strength; melting point is 449°F, boiling point is 4384°F, yielding the longest molten-state range for any common metal; specific gravity is 7.28. Its principal use is as a coating on steel in tin plate, also as a constituent in alloys.

Titanium (Ti). Element no. 22 of the periodic system; atomic weight 47.90; melting point is about 3270°F; boiling point is over 5430°F; specific gravity is 4.5. Bright white metal, very malleable and ductile when exceedingly pure. Its principal function is as an alloy in making steel. It fixes carbon in inert particles to reduce martensitic hardness and hardenability in medium chromium steels, prevents formation of austenite in high-chromium steels, and prevents localized depletion of chromium in stainless steel during long heating. It now is finding application in its own right because of its high strength and good corrosion resistance.

Titanium-based superalloys. Lightweight, corrosive-resistant alloys suitable for high temperatures. These alloys are very practical for airplane parts. Titanium alloys can be blended with aluminum, iron, vanadium, silicon, cobalt, tantalum, zirconium, and manganese.

Tolerance limit. The permissible deviation from the desired value.

Tolerances. A customer's specifications can refer to dimensions or to the chemical properties of the steel ordered. The tolerance measures the allowable difference in product specifications between what a customer orders and what the steel company delivers. There is no standard tolerance because each customer maintains its own variance objective. Tolerances are given as the specification, plus or minus an error factor; the smaller the range, the higher the cost.

Ton. Unit of measure for stainless steel scrap and iron ore:

Gross ton: 2240 pounds.
Long (net) ton: 2240 pounds.
Short (net) ton: 2000 pounds. Normal unit of statistical raw material input and stainless steel output in the United States.
Metric ton: 1000 kg (2204.6 lb or 1.102 short tons).

Torsion. The twisting action resulting in shear stresses and strains.

Toughness. An indication of a steel's capacity to absorb energy, particularly in the presence of a notch or a crack.

Trace element. Extremely small quantity of an element, usually too small to determine quantitatively.

Transformation. A constitutional change in a solid metal, such as the change from gamma to alpha iron or the formation of pearlite from austenite.

Transformation ranges (transformation temperature ranges). Those ranges of temperature within which austenite forms during heating and transforms during cooling. The two ranges are distinct, sometimes overlapping but never coinciding. The limiting temperatures of the ranges depend on the composition of the alloy and on the rate of change of temperature, particularly during cooling.

Tubing. When referring to oil country tubular goods, tubing is a separate pipe used within the casing to conduct the oil or gas to the surface. Depending on conditions and well life, tubing may have to be replaced during the operational life of a well.

Tungsten (W). Element no. 74 of the periodic system; atomic weight 183.92. Gray metal of high tensile strength, ductile and malleable when specially handled. It is immune to atmospheric influences and most acids but not to strong alkalis. The metal is used as filament and in thin sheet form in incandescent bulbs and radio tubes. Forms hard abrasion-resistant particles in tool steels and promotes hardness and strength at elevated temperatures.

Tungsten carbide. Compound of tungsten and carbon, of composition varying between WC and W_2C; imbedded in matrix of soft metal, such as cobalt, extensively used for sintered carbide tools.

Tungsten inclusion. An inclusion of tungsten from the electrode in TIG welding.

Tunnel furnace. Type of furnace whereby stock to be heated is placed on cars pushed or pulled slowly through the furnace.

Twist. A winding departure from flatness.

Two-coat system. The combination of a prime coat and a finish coat into a specified paint film. A typical 1 ml, two-coat system will have about 0.2 ml of primer coat and about 0.8 ml of finish coat.

Ultimate strength. The maximum conventional stress—tensile, compressive, or shear—that a material can withstand.

Universal mill. A rolling mill in which rolls with a vertical axis roll the edges of the metal stock through the horizontal rolls between some of the passes.

Vacuum degassing. An advanced steel refining facility that removes oxygen, hydrogen, and nitrogen under low pressures (in a vacuum) to produce ultra-low-carbon steel for demanding electrical and automotive applications. Normally performed in the ladle, the removal of dissolved gases results in cleaner, higher-quality, purer steel.

Vacuum oxygen decarburization (VOD). A refinement of stainless steel that reduces carbon content. Molten, unrefined stainless steel is heated and stirred by an electrical current while oxygen enters from the top. Many undesirable gases escape from the stainless steel and are evacuated by a vacuum pump. Alloys and other additives are then mixed in to refine the molten stainless steel further.

Vanadium (V). A gray metal normally used as an alloying agent for iron and stainless steel. It is also used as a strengthener of titanium-based alloys.

Vickers hardness (test). Standard method for measuring the hardness of metals, particularly those with extremely hard surfaces: The surface is subjected to a standard pressure for a standard length of time by means of a pyramid-shaped diamond. The diagonal of the resulting indention is measured under a microscope and the Vickers hardness value read from a conversion table.

Weathering steel. A steel using alloying elements such as copper, chromium, silicon, or nickel to enhance its resistance to atmospheric corrosion. (USS COR-TEN)

Welding. A process used to join metals by the application of heat. Fusion welding, which includes gas, arc, and resistance welding, requires that the parent metals be melted. This distinguishes fusion welding from brazing. In pressure welding, joining is accomplished by the use of heat and pressure without melting. The parts that are being welded are pressed together and heated simultaneously, so that recrystallization occurs across the interface.

Wet-film thickness (WFT). The thickness of the paint film immediately after coating and prior to curing. The required wet-film thickness depends on the proportion of solids and solvents in the liquid paint for producing the appropriate dry-film thickness.

Width. The lateral dimension of rolled steel, as opposed to the length or the gauge (thickness). If the width of the steel strip is not controlled during rolling, the edges must be trimmed.

Wire. A cold finished, stainless steel product (normally in coils) that is round, square, octagon, hexagon, or flat and less than $^{3}/_{16}$ in. in thickness.

Workability. The characteristic or group of characteristics that determines the ease of forming a metal into desired shapes.

Work hardening. Increase in resistance to deformation (i.e., in hardness) produced by cold working.

Wrought iron. Iron containing only a very small amount of other elements but containing 1–3% by weight of slag in the form of particles elongated in one direction, giving the iron a characteristic grain. It is more rust resistant than steel and welds more easily.

X-rays. Light rays, excited usually by the impact of cathode rays on matter, that have wavelengths between about 10.6 cm and 10.9 cm.

Yield point. The load per unit of original cross-section at which, in soft steel, a marked increase in deformation occurs without increase in load.

Yield strength. The stress beyond which stainless steel undergoes important permanent flow; commonly specified as that stress producing a 0.2% offset from the linear portion of the stress-strain curve.

Young's modulus. The coefficient of elasticity of stretching. For a stretched wire, Young's modulus is the ratio of the stretching force per unit of cross-sectional area to the elongation per unit of length. The values of Young's modulus for metals are on the order of 10/12 dynes per square cm.

Zinc (Zn). Element no. 30 of the periodic system; atomic weight 65.38. Blue-white metal; when pure, malleable and ductile even at ordinary temperatures; melting point is 787°F; boiling point is 1665°F; specific gravity is 7.14. It can be electrodeposited; it is extensively used as a coating for steel; and sheet zinc finds many outlets, such as dry batteries. Zinc-base alloys are of great importance in die casting. Its most important alloy is brass.

Zirconium (Zr). Element no. 40 of the periodic system; atomic weight 91.22. Specific gravity is 6.5, and melting point is at about 3200° ± 1300°F. Because of its great affinity for oxygen and ability to combine readily with nitrogen and sulfur, it is used as a deoxidizer and scavenger in steel making. It is used as an alloy with nickel for cutting tools and in copper alloys.

2. WELDING GLOSSARY

To specify the materials of construction and the fabrication techniques necessary to complete a process plant, the piping engineer must be familiar with welding terminology. If in any doubt, the piping engineer must refer to a specialist, welding engineer or metallurgist, for advice. This glossary contains welding terms commonly used in the oil and gas industry.

Actual throat thickness. The perpendicular distance between two lines each parallel to a line joining the outer toes, one line tangent at the weld face and the other through the furthermost point of fusion penetration.

Air-arc cutting. Thermal cutting using an arc to melt the metal and a stream of air to remove the molten metal to enable completion of a cut.

All position. A gas welding technique in which the flame is rightward welding, meaning the weld is started on the left and travels to the right.

All-weld test piece. A piece of metal consisting of one or more beads or runs fused together for test purposes. It may include portions of parent metal.

Arc blow. A lengthening or deflection of a dc welding arc caused by the interaction of magnetic fields set up in the work and arc or cables.

Arc fan. A fan-shaped flame associated with the atomic-hydrogen arc.

Arc voltage. The voltage between electrodes or between an electrode and the work, measured at a point as near as practical to the work.

Atomic-hydrogen welding. Arc welding in which molecular hydrogen passes through an arc between two tungsten or suitable electrodes, is changed to its atomic form, then combines to supply the heat for welding.

Backfire. Retrogression of the flame into the blowpipe neck or body with rapid self-extinction.

Backing bar. A piece of metal or other material placed at the root of the weld.

Backing strip. A strip of metal placed at the root of a weld.

Back-step sequence. A welding sequence in which short lengths of weld are deposited adjacent to each other.

Blowhole. A cavity, generally over 1.6 mm in diameter, formed by entrapped gas during solidification of molten metal.

Blowpipe. A device for mixing and burning gases to produce a flame for welding, brazing, bronze welding, cutting, heating, and similar operations.

Burn back. Fusing an electrode wire to the current contact tube by a sudden lengthening of the arc in any form of automatic or semi-automatic metal-arc welding using a bare electrode.

Burn-off rate. The linear rate of consumption of a consumable electrode.

Burn through. A localized collapse of the molten pool due to penetration of the workpiece.

Carbon-arc welding. Arc welding using a carbon electrode.

Chain intermittent weld. An intermittent weld on each side of a joint (usually fillet welds in T and lap joints) arranged so that the welds lie opposite one another along the joint.

Concave fillet weld. A fillet weld in which the weld face curves inward.

Cone. The most luminous part of a flame, which is adjacent to the nozzle orifice.

Continuous weld. A weld that extends along the entire length of a joint.

Convex fillet weld. A fillet weld in which the weld face curves outward.

CO_2 flux welding. Metal-arc welding using a flux-coated or flux-containing electrode that is deposited under a shield of carbon dioxide.

CO_2 welding. Metal-arc welding in which a bare wire electrode is used; the arc and molten pool are shielded with carbon dioxide.

Coupon plate. A test piece made by adding plates to the end of a joint to extend the weld for test purposes.

Crack. A longitudinal discontinuity produced by a fracture. Cracks may be longitudinal, transverse, edge, crater, center line, and fusion zone and located on the weld metal or parent metal.

Crater pipe. A depression caused by shrinkage at the end of a run where the source of heat was removed.

Cutting electrode. An electrode with a covering that aids the production of such an arc that molten metal is blown away to produce a groove or cut in the work.

Cutting oxygen. Oxygen used at a pressure suitable for cutting metal.

Deseaming. The removal of the surface defects from ingots, blooms, billets, and slabs by manual thermal cutting.

Dip transfer. A method of metal-arc welding in which fused particles of the electrode wire in contact with the molten pool are detached from the electrode in rapid succession by the short circuit current that develops every time the wire touches the molten pool.

Drag. The projected distance between the two ends of a *drag line.*

Drag lines. Serrations left on the face of a cut made by thermal cutting.

Electron-beam cutting. Thermal cutting in a vacuum by melting and vaporizing a narrow section of the metal by the impact of a focused beam of electrons.

Excess penetration bead. Excessive metal protruding through the root of a fusion weld made from one side only.

Feather. The carbon-rich zone, visible in a flame, that extends around and beyond the cone when there is an excess of carbonaceous gas.

Fillet weld. A fusion weld, other than a butt, edge, or fusion spot weld, which is approximately triangular in transverse cross-section.

Flame cutting. Oxygen cutting in which the appropriate part of the material to be cut is raised to ignition temperature by an oxy-fuel gas flame.

Flame snap-out. Retrogression of the flame beyond the blowpipe body into the hose, with a possible subsequent explosion.

Flame washing. A method of surface shaping and dressing metal by flame cutting using a nozzle designed to produce a suitably shaped cutting oxygen stream.

Flashback arrestor. A safety device fitted in the oxygen and fuel gas system to prevent any flashback reaching the gas supplies.

Floating head. A blowpipe holder on a flame cutting machine that, through a suitable linkage, is designed to follow the contour of the surface of the plate, thereby enabling the correct nozzle-to-workpiece distance to be maintained.

Free bend test. A bend test made without using a former.

Fusion penetration. In fusion welding, the depth to which the parent metal has been fused.

Fusion zone. The part of the parent metal melted into the weld metal.

Gas economizer. An auxiliary device designed for temporarily cutting off the supply of gas to the welding equipment except the supply to a pilot jet where fitted.

Gas envelope. The gas surrounding the inner cone of an oxy-gas flame.

Gas pore. A cavity, generally under 1.6 mm in diameter, formed by gas entrapped during solidification of molten metal.

Gas regulator. An attachment to a gas cylinder or pipeline for reducing and regulating the gas pressure to the working pressure required.

Guided bend test. A bend test made by bending the specimen around a specified former.

Heat-affected zone. The part of the parent metal metallurgically affected by the heat of welding or thermal cutting but not melted. Also known as the "zone of thermal disturbance."

Hose protector. A small nonreturn valve fitted to the blowpipe end of a hose to resist the retrogressive force of a flashback.

Included angle. The angle between the planes of the fusion faces of parts to be welded.

Inclusion. Slag or other foreign matter entrapped during welding. The defect is usually more irregular in shape than a gas pore.

Incompletely filled groove. A continuous or intermittent channel in the surface of a weld, running along its length, due to insufficient weld metal. The channel may be along the center or one or both edges of the weld.

Incomplete root penetration. Failure of the weld metal to extend into the root of a joint.

Intermittent weld. A series of welds at intervals along a joint.

Kerf. The void left after metal has been removed by thermal cutting.

Lack of fusion. Lack of union in a weld (between the weld metal and parent metal, parent metal and parent metal, or weld metal and weld metal).

Leftward welding. A gas welding technique in which the flame is started on the right and travels to the left (forward welding).

Leg. The width of a fusion face in a fillet weld.

Metal-arc cutting. Thermal cutting by melting, using the heat of an arc between a metal electrode and the metal to be cut.

Metal-arc welding. Arc welding using a consumable electrode.

Metal transfer. The transfer of metal across the arc from a consumable electrode to the molten pool.

MIG (metal inert gas) **welding.** Inert-gas welding using a consumable electrode.

Multistage regulator. A gas regulator in which the gas pressure is reduced to the working pressure in more than one stage.

Nick-break test. A fracture test in which a specimen is broken from a notch cut at a predetermined position where the interior of the weld is to be examined.

Open arc welding. Arc welding in which the arc is visible.

Open circuit voltage. In a welding plant ready for welding, the voltage between two output terminals carrying no current.

Overlap. An imperfection at a toe or root of a weld caused by metal flowing onto the surface of the parent metal without fusing it.

Oxygen-arc cutting. Thermal cutting in which the ignition temperature is produced by an electric arc, and cutting oxygen is conveyed through the center of an electrode that is consumed in the process.

Oxygen lance. A steel tube, consumed during cutting, through which cutting oxygen passes, for cutting or boring holes.

Oxygen lancing. Thermal cutting in which an oxygen lance is used.

Packed lance. An oxygen lance with steel rods or wires.

Penetration bead. Weld metal protruding through the root of a fusion weld made from one side only.

Plug weld. A weld made by filling a hole in one component of a workpiece to join it to the surface of an overlapping component, exposed through the hole.

Porosity. The presence of gas pores.

Powder cutting. Oxygen cutting in which powder is injected into the cutting oxygen stream to assist the cutting action.

Powder lance. An oxygen lance in which powder is mixed with the oxygen stream.

Preheating oxygen. Oxygen used at a suitable pressure in conjunction with fuel gas to raise the ignition temperature of the metal to be cut.

Residual welding stress. Stress remaining in a metal part or structure as a result of welding.

Reverse bend test. A bend test in which the side other than that specified for a face bend test is in tension.

Rightward welding. A gas welding technique in which the flame is started on the left and travels to the right (backward welding).

Root (of weld). The zone on the side of the first run farthest from the welder.

Root face. The portion of a fusion face at the root that is not beveled or grooved.

Run-off-plate(s). A piece, or pieces, of metal so placed as to enable the full section of weld to be obtained at the end of the joint.

Run-on-plate(s). A piece, or pieces, of metal so placed as to enable the full section of weld metal to be obtained at the beginning of a joint.

Scarfing. The removal of the surface defects from ingots, blooms, billets, and slabs by a flame cutting machine.

Sealing run. The final run deposited on the root side of a fusion (backing run).

Seal weld. A weld, not a strength weld, used to make a sealing weld.

Shrinkage groove. A shallow groove caused by contraction of the metal along each side of a penetration bead.

Side bend test. A bend test in which the face of a transverse section of the weld is in tension.

Skip sequence. A welding sequence in which short lengths of run are skip welded.

Slag trap. A configuration in a joint or joint preparation designed to entrap slag.

Slot lap joint. A joint between two overlapping components made by depositing a fillet weld around the periphery of a hole in one component to join it to the other component, exposed through the hole.

Spray transfer. Metal transfer that takes place as globules of diameter substantially larger than that of the consumable electrode from which they are transferred.

Stack cutting. The thermal cutting of a stack of plates, usually clamped together.

Staggered intermittent weld. An intermittent weld on each side of a joint (usually fillet welds in T and lap joints) arranged so that the welds on one side lie opposite the spaces on the other side along the joint.

Striking voltage. The minimum voltage at which any specified arc may be initiated.

Submerged-arc welding. Metal-arc welding in which a bare wire electrode or electrodes are used; the arc or arcs are enveloped in a flux, some of which fuses to form a removable covering of slag on the weld.

Surface-fusion welding. Gas welding in which a carburizing flame is used to melt the surface of the parent metal, which then unites with the metal from a suitable filler rod.

Sustained backfire. Retrogression of the flame into the blowpipe neck or body of the flame remaining alight. Note: This manifests itself either as "popping" or "squealing," with a small pointed flame issuing from the nozzle orifice or as a rapid series of minor explosions inside.

Test piece. Components welded together in accordance with a specified welding procedure or a portion of a welded joint detached from a structure for testing.

Test specimen. A portion detached for a test piece and prepared as a test coupon.

Thermal cutting. The parting or shaping of materials by the application of heat with or without a stream of cutting oxygen.

TIG (tungsten inert gas) **welding.** Inert-gas welding using a nonconsumable electrode (inert-gas tungsten-arc welding).

Toe. The boundary between a weld face and the parent metal or between weld faces.

Tongue-bend test specimen. A portion so cut in two straight lengths of pipe joined by a butt weld to produce a tongue containing a portion of the weld. The cuts are made so that the tongue is parallel to the axis of the pipes, and the weld is tested by bending the tongue around.

Touch welding. Metal-arc welding using a covered electrode, the covering of which is kept in contact with the parent metal during welding.

Tungsten inclusion. An inclusion of tungsten from the electrode in TIG welding.

Two-stage regulator. A gas regulator in which the gas pressure is reduced to the working pressure in two stages.

Undercut. An irregular groove at a toe of a run in the parent metal or in previously deposited weld metal due to welding.

Weld junction. The boundary between the fusion zone and the heat affected zone.

Welding procedure. A specified course of action followed in welding, including the list of materials and, where necessary, tools to be used.

Welding sequence. The order and direction in which joints, welds, or runs are made.

Welding technique. The manner in which the operator manipulates an electrode, a blowpipe, or a similar appliance.

Worm hole. An elongated or tubular cavity formed from gas entrapped during the solidification of molten metal.

3. REFINERY GLOSSARY

As all process industry projects require creating piping classes and specifications for the transport of some form of the process, it is essential that the fundamental terminology used by process engineers be comprehended. This understanding allows the piping engineer to select the most efficient material and piping components necessary for the process plant. Process engineers create the process philosophy, but it is essential that the piping engineer fully understand their requirements. This glossary contains internationally and commonly used words and terms in the refining of petroleum products.

Absorption. The process by which one substance attracts and encompasses another, forming a homogeneous mixture. Oil absorbs natural gasoline from wet natural gas. A caustic absorbs hydrogen sulfide from hydrocarbon vapors.

Accumulator. A temporary storage tank for liquids and vapors.

Acid treatment. A process in which unfinished petroleum products, such as gasoline, kerosene, and lubricating oils, are treated with sulfuric acid to improve the color, odor, and other characteristics.

Additive. A substance added to petroleum products to impart some desirable property.

Adsorption. The process by which one substance attracts another, forming a physical or chemical bond at its surface.

Agitator. A cone-bottom tank for treating oils equipped with air or gas spargers for mixing.

Air blowing. A process for raising the softening point of an asphalt by reaction with air at elevated temperatures.

Air fin coolers. A cooling device with radiatorlike fins used to cool or condense hot hydrocarbons. Also known as "fin fans."

Alkylate. The product of an alkylation process or to perform that process.

Alkylation. The process of combining an olefin with an isoparaffin to form an isoparaffin of higher molecular weight. Also, combining an olefin with an aromatic to form alkyl-benzene.

API. American Petroleum Institute.

API gravity. A special gravity scale adopted by the API to express the gravities of petroleum products.

Aromatic. Unsaturated ring-structured hydrocarbon molecule.

ASME. American Society of Mechanical Engineers.

Asphalt. Black to dark brown solid or semisolid bituminous material, which gradually liquefies when heated, produced from distillation residues of crude petroleum or occurring naturally.

Asphaltenes. A principal component of asphalt, it is the black or brown solid material precipitated from an asphalt with normal pentane. It is an arbitrary fraction defined by the method of analysis. Other arbitrary fractions of asphalt are oils and resins.

Assay. The tabulated results of a comprehensive laboratory analysis of crude oil.

ASTM (American Society of Testing and Materials). An organization that sets standards for the testing of industrial products.

Atmospheric tower. A distillation unit that operates at atmospheric pressure.

Barrel. The standard unit of measurement in the petroleum industry. It contains 42 U.S. standard gallons, 35 imperial gallons. (BPD is barrels per day.)

Base oil. A finished petroleum stock, which is blended with other materials to make saleable products.

Battery limit. The perimeter of a process facility or unit.

Benzene. An unsaturated, basic aromatic compound.

Bitumen. Hydrocarbon material of natural or pyrogenous origin, or combination of both, accompanied by nonmetallic derivatives, which may be gaseous, liquid, semisolid, or solid, and is completely soluble in carbon disulphide (ASTM D 8-63).

Blanket gas. Gas introduced above a liquid in a vessel to keep out air to prevent oxidation of the material or prevent forming explosive mixtures.

Blending. Mixing two or more materials together.

Blind. A steel plate inserted between a pair of flanges to prevent flow through a line.

Block valve. A valve used to isolate equipment or piping systems.

Bloom. The color of an oil observed by reflected light.

Blowdown. Withdrawal of water from boilers and cooling towers to prevent buildup of solids.

Blower. Equipment used to move large volumes of gas against low-pressure heads.

Boiling range. The temperature range, usually at atmospheric pressure, at which the boiling, or distillation of a hydrocarbon liquid, commences and finishes.

Bottoms. Residue remaining in a still after distillation. "Tank bottoms" refers to water and sediment in the tank.

Briddle. A screwed or socket-weld assembly used to measure the liquid level in a horizontal or vertical vessel.

Bright stock. High viscosity, refined, and dewaxed lubrication oil base stock, usually produced by suitable treatment of petroleum residues.

BS and W. Bottoms sediment and water measured in a crude oil by centrifuging a sample, sometimes abbreviated "S and W."

Bubble tower. A fractionating or distillation tower in which the rising vapors pass through layers of condensate, bubbling under caps on a series of trays at differing elevations.

By-products. Useful materials recovered incidental to the principal objective of refining petroleum.

Catalyst. Material that promotes a chemical reaction but remains unchanged itself or can be regenerated to its original form.

Catalytic cracking. A process in which large molecules are broken into smaller molecules by the use of heat, pressure, and catalyst.

Catalytic desulphurization. A process in which the sulfur content of petroleum is reduced, usually by conversion to H_2S (Hydrogen Sulphide), using a catalyst with or without the presence of added hydrogen.

Catalytic reforming. A process in which naphthenes are converted to aromatics by removal of hydrogen in the presence of a catalyst.

Caustic wash. A process in which the distillate is treated with sodium hydroxide (a caustic) to remove acidic contaminants that cause poor odor and lack of stability.

Chromatograph. An apparatus for analyzing mixtures of compounds by separating them into individual components, which can be identified by color or other means.

Clay. Granular or finely divided mineral material used for treating petroleum. This is a general term including fuller's earth, bauxite, bentonite, and montmorillonite.

Coke. A high-carbon residue that remains after the destructive distillation of petroleum product.

Coking. A thermal process used for converting and upgrading heavy residual into lighter products. This leaves a residue of carbon deposits in the process equipment.

Compounding. Mixing additives with oils, particularly lubes, to impart oxidation resistance, rust resistance, or detergency.

Condensate. Liquid condensed from the vapors leaving the top of a distillation column.

Condenser. A heat treatment item of process equipment that cools and condenses by removing heat via a cooling medium, which could be water or lower-temperature hydrocarbon streams.

Control valve. A valve, usually of a globe valve pattern, used to automatically control the flow of a fluid and pressure through a piping system.

Cracked naphtha. The crude, low-boiling product of the thermal cracking process, from which gasoline is made by distillation.

Cracked tar. Residue from the thermal cracking process.

Cracking. The breaking up of heavy-molecular-weight hydrocarbons into lighter hydrocarbon molecules by applying heat and pressure, with or without the use of catalysts.

Crude. A short name for raw or unrefined petroleum, crude oil.

Crude assay. The procedure for determining the distillation characteristics of crude oil.

Cut. A fraction, a part of the whole, such as the gasoline cut from raw crude oil.

Cycle gas oil. Cracked gas oil returned to the cracking unit.

Cycle stock. Material taken from a later stage of a process and recharged to the process at some earlier stage. Light cycle oil, a fraction of the synthetic crude produced in the catalytic cracking process, returned to the reactor is an example.

Cylinder stock. Oil used for lubrication of steam cylinders, usually a high viscosity distillate.

Debutanizer. A fractionating column used to remove butane and lighter components from liquid streams.

De-ethanizer. A fractionating column designed to remove ethane and gases from heavier hydrocarbons.

Dehydrogeneration. A reaction in which hydrogen atoms are eliminated from a molecule. Used to convert ethane, propane, and butane into olefins (ethylenes, propylenes, and butanes).

Demulsification. Process of breaking up or separating an emulsion into its components.

Demulsifier. An additive that promotes demulsification.

Deoiling. The process of making an oil-free wax from a waxy stock, by chilling a mixture of solvent and feed to crystallize part of the wax and separating the wax from the waxy-oil-solvent by filtering or centrifuging.

Depentanizer. A fractionating column used to remove pentane and lighter fractions from hydrocarbon streams.

Depropanizer. A fractionating column used to remove propane and lighter fractions from hydrocarbon streams.

Desalting. The process of removing salt from crude by emulsification with water, then breaking the emulsion and separating phases.

Desulfurization. A chemical treatment to remove sulfur or sulfur compounds from hydrocarbons.

Dewaxing. The process of making a wax-free oil from a waxy stock, by chilling a mixture of solvent and feed to a low temperature to crystallize all the wax and separating the wax cake (slack wax) by filtering or centrifuging.

Dew point. The temperature at which a mixture of gases reaches saturation conditions for one or more of the components and condensation begins.

Distillate. The products of distillation formed by condensing vapors.

Downcomer, downspout. That part of a distillation column tray in which liquid is separated from the froth and flows by gravity to a tray below.

Downflow. The process stream that flows from top to bottom.

Draw, draw off. A pipe connection (usually valved) to a vessel through which liquid can flow. For example, a side cut draw on a distillation column or a water draw on a tank.

Dry gas. A hydrocarbon gas consisting mainly of methane and ethane and containing no recoverable amounts of butane or heavier hydrocarbons.

End point (EP). The highest temperature indicated by the thermometer in a distillation test of an oil.

Extraction. The process of separating a material by means of a partially miscible solvent into a fraction soluble in the solvent and a fraction relatively insoluble.

Feedstock. Stock used to charge or feed the processing unit.

Filter. A basket-type strainer used to collect solid waste in the piping system.

Fixed gas. Gas that does not condense under the pressure and temperature conditions available in a process.

Flashing. The process of separating products by reducing the pressure on a hot oil as it enters a vessel. The light fractions vaporize (flash off) while the liquid drops to the bottom.

Flash point. The lowest temperature at which an oil gives off vapor in sufficient quantity to burn momentarily on the approach of a flame or spark.

Flash tower. A vessel used to separate liquid and vapor in a flash distillation process.

Flood, flooding. A spewing or heaving of liquid along with vapor out of the top of a distillation column. It is caused by excessive liquid or vapor flow in the column.

Flux, flux oil. A liquid residuum from asphaltic crude.

Fraction. A portion of fractional distillation that has a restricted boiling range.

Fractionating column. Process unit that separates various fractions of petroleum by simple distillation, at different levels.

Fuel gas. Light gas by-products from refining operations that are used to fuel process furnaces.

Fuel oil. A general term applied to any oil used to produce power and heat. For a description of classes of fuels, see ASTM D 288.

Furnace. A term applied to oil- and gas-fired heaters used in refinery processes.

Gas blanket. An atmosphere of inert gas applied to the vapor space of a vessel to prevent oxidation of its contents and explosive mixtures from forming.

Gasoline, cracked. The principle product of catalytic cracking and thermal cracking processes.

Gasoline, natural. Liquid product recovered from wet natural gas by absorption, compression, or refrigeration.

Gasoline, polymer. Product of polymerization of normally gaseous olefins to hydrocarbons boiling in the gasoline range.

Gasoline, straight run. A product obtained by distillation of petroleum.

Header. A common line to which two or more lines are joined.

Heater. The furnace and tube arrangement that furnishes heat for a process.

Heat exchanger. A piece of equipment that transfers heat from a flowing stream at one temperature to another stream flowing at another temperature. Heat is transferred between the liquid or gases through a tubular wall.

High-line, high-pressure gas. High-pressure (100 psi) gas from the cracking unit distillate drums compressed and combined with low-line gas as gas absorption feedstock.

Hydrocarbon. Compounds of carbon and hydrogen. Crude oil is a complex mixture of hydrocarbon compounds.

Hydrocracking. A process in which large molecules are broken into smaller molecules by heat, pressure, catalyst, and hydrogen.

Hydrodesulfurization. A catalytic process to remove sulfur from petroleum fractions in the presence of hydrogen.

Hydrofining. A process for treating petroleum with hydrogen in the presence of a catalyst under relatively mild conditions of temperature and pressure. Some hydrocracking may occur under more severe conditions.

Hydrofinishing. A mild hydrofining process used particularly to replace or supplement clay treatment of lube oils and waxes.

Hydroforming. Catalytic reforming of naphtha at elevated temperatures and moderate temperatures in the presence of hydrogen to form high-octane BTX (benzene, toluene and xylenes) aromatics for motor fuel and chemical manufacture.

Hydrogenation. A hydrogen treating process in which hydrogen is added to unsaturated molecules. Olefins are converted to paraffins, aromatics are converted to naphthenes, oxygen is converted to water, nitrogen is converted to ammonia, and sulfur is converted to H_2S.

Inhibitor. An additive used to prevent or delay an undesirable change in the quality of the process or the individual process equipment.

Initial boiling point (init or IBP). The temperature in a laboratory distillation test indicated by the thermometer at which the first drop of distillate falls from the condenser.

Isomerization. A process for rearranging the atoms in a molecule so that the product has the same empirical formula but a different structure, such as converting normal butane to isobutane.

Knock-out drum. A vessel in which suspended liquid is separated from the gas or vapor.

Lean oil. Absorption oil from which gasoline fractions have been removed, such as the oil leaving the stripper in a natural gasoline plant.

LNG. Abbreviation for liquefied natural gas.

Low-line or low-pressure gas. Low-pressure (5 psi) gas from atmospheric or vacuum distillation recovery systems collected in the gas plant for compression to a higher pressure.

LPG. Abbreviation for liquefied petroleum gas.

Lube. Short for lubricating oil.

Lube distillate. High-boiling, 700–1000°F range, petroleum distillate used for manufacture of lubricating oils.

Mercaptans. Sulfur compounds occurring naturally in some crudes and formed in cracking operations. They are foul-smelling substances of the formula RSH, where R is an alkyl group.

Naphtha. A general term applied to the lower boiling fractions of petroleum, usually below 400°F.

Naphthene. A hydrocarbon containing at least one ring structure and saturated with respect to hydrogen.

Naphthenic acids. The organic acids occurring naturally in petroleum.

Natural gas. Naturally occurring light hydrocarbons varying in composition from mostly methane to a mixture ranging from methane through hexanes.

Neutral oil. Lubricating oil base stock of low and medium viscosity, generally light in color and produced from lube oil distillation or distilled from a refined and dewaxed residuum.

Octane number or rating. A measure of the antiknock qualities of gasoline. Isooctane is rated at 100. Normal heptane is rated at 0. A 50–50 mixture of these two would be rated at 50.

Olefins. Open-chain hydrocarbons containing one or more double bonds.

Paraffins. Open-chain hydrocarbons saturated with respect to hydrogen.

Paraffin wax, petroleum wax. Hydrocarbons of molecular weight higher than 250 and boiling above 600°F that are solid at room temperature.

Waxes are mostly straight hydrocarbons with a small amount of branching. They may also contain naphthene and aromatic rings.

Preheater. An exchanger used to heat hydrocarbons before they are fed to a process unit.

Polymerization. The process of combining two or more molecules to form a single large molecule.

Raffinate. The refined oil produced in solvent extraction processes.

Reactor. A vessel in which a chemical reaction takes place.

Reboiler. An auxiliary unit used to supply additional heat to the lower portion of a fractionating tower.

Rectification. Fractional distillation referring to the removal of low-boiling hydrocarbons from gasoline to adjust its vapor pressure.

Recycling gas. High-hydrogen-content gas returned to a unit for reprocessing.

Reflux. The portion of the distillate returned to the fractionating column to assist in achieving better separation into desired fractions.

Reformate. An upgraded naphtha resulting from catalytic or thermal reforming.

Reforming. A process of cracking gasoline to increase its octane number.

Regeneration. The reactivation of the catalyst in a process unit.

Resins. One of the three main constituents of asphalt, characterized by absorption on clay.

Rich oil. Absorption oil containing dissolved gasoline fractions.

Riser. A pipe that allows vapor or liquid to flow upward in a process vessel.

Rundown tanks. Receiving tanks for products of distillation, also called "pans."

Scrubbing. Purification of a gas by washing it in a tower or agitator.

Seal oil, sealing oil. Any oil used to pump packing glands to keep packing cool and prevent leakage of stock being processed.

Sour. Foul smelling. Describing a petroleum fraction having a bad odor, usually caused by mercaptans.

Sour gas. Natural gas that contains corrosive, sulfur-bearing compounds such as hydrogen sulfide and mercaptans.

Stabilization. A process to separate the gaseous and more volatile liquid hydrocarbons from crude petroleum, thereby stablizing the product for safer handling or storage.

Straight run. A material produced directly from crude oil by distillation.

Stripping. The removal, by steam induced vaporization or flash evaporation, of the more volatile components from a cut or fraction.

Sulfurization. Combining sulfur compounds with petroleum lubricants.

Sweet. Having a good odor.

Sweetening. A process of removing mercaptans or converting them to disulfides.

Tail gas. The lightest hydrocarbon gas released from a refining process.

Tar. Petroleum residuum, either straight run or cracked.

Thermal cracking. The process of breaking down large molecules into smaller ones by heat and pressure.

Thinners. Narrow boiling fractions in the gasoline and kerosene boiling range (200–450°F) used as solvents and dilutents in the paint and varnish industry.

Topping. The process of distillation in which only a fraction of the distillable portion of the feed is removed, such as topped crude, and atmospheric column bottoms.

Treating. Refining petroleum with chemicals.

Turnaround. A planned shutdown of an entire process or section of a refinery for major maintenance.

Unsaturates, unsaturated hydrocarbons. Hydrocarbons deficient in hydrogen, including olefins and aromatics.

Vacuum distillation. Distillation at reduced pressure (below atmospheric pressure), with or without the use of steam.

Vapor. Gaseous substance that can be at least partly condensed by cooling or compression.

Vapor binding. The formation of vapor in a liquid line causing an interruption of liquid flow.

Vapor lock. Vapor binding in the gasoline line leading to an engine.

Virgin. A descriptive term applied to products produced directly from crude to distinguish them from similar products made from cracking processes.

Visbreaking. A mild thermal cracking process.

Viscosity. The resistance of a fluid to flow when a force is exerted on it.

Viscosity-gravity constant. A useful function for characterizing viscous fractions of petroleum, see ASTM D 2501-67.

Viscosity index (VI). An empirical number indicating the effect of change in temperature on the viscosity of an oil. A high viscosity index signifies a relatively small change of viscosity with temperature, see ASTM D 2270-64.

Wax, petroleum wax. A product separated from petroleum, which is solid or semi-solid at 77°F and consists essentially of a mixture of saturated hydrocarbons.

Weathering. The process of vaporization, which occurs when the crude of petroleum product is allowed to stand in an open vessel, also the deterioration of a material resulting from exposure to the atmosphere.

Wet gas. Natural gas, or gas produced by some refinery process, that contains recoverable gasoline fractions.

4. PIPING ABBREVIATIONS

#	Lbs or class (ASME)
BB	Bolted bonnet
BW	Butt-welding ends
CE	Carbon equivalent
CI	Gray cast iron (GG25)
CS	Carbon steel

DN	Nominal diameter
EFW	Electric fusion welded
ERW	Electric resistance welded
FB	Full bore
FF	Flat face
F/F	Face to face
HB	Hardness, Brinnel
HRB	Hardness, Rockwell B method
HRC	Hardness, Rockwell C method
ID	Inside diameter
IS&Y	Inside screw and yoke
LPI(E)	Liquid penetrant inspection (examination)
MPI(E)	Magnetic particle inspection (examination)
MTC	Material test certificate
ND	Nominal diameter
NDT(E)	Nondestructive test (examination)
NPS	Nominal pipe size
OD	Outside diameter
OS&Y	Outside screw and yoke
PE	Plain ends
PMI	Positive material identification
PN	Nominal pressure
PWHT	Postweld heat treatment
QT	Quench and tempered
RB	Reduced bore
RF	Raised face
RTJ	Ring-type joint
SAW	Submerged arc welded
SCH	Schedule (wall thickness)
SG	Spheroidal graphite cast iron
SMLS	Seamless
SR	Supplementary requirement
SS	Stainless steel
SW	Socket weld
TPI	Third party inspection
UNS	Unified numbering system
US	Ultrasonic test
WT	Wall thickness
XS	Extra strong (wall thickness)
XXS	Extra extra strong (wall thickness)

5. ELASTOMER AND POLYMER GLOSSARY

This glossary is a list of terms commonly used in the production and application of elastomers and polymers.

Abrasion. The wearing away of a surface by mechanical action, such as rubbing, scraping, or erosion.

Accelerator. A substance that hastens the vulcanization of an elastomer, causing it to take place in a shorter time or at a lower temperature.

Acrylic. A polymer for which resistance to air and hot oil at temperatures above 30°F are required.

Adhere. To cling or stick together.

Adhesion. The tendency to bond or cling to a contact surface.

Aging. To undergo changes in physical properties with age or lapse of time.

Air curing. Vulcanization of a rubber product in air, as distinguished from in a press or steam vulcanizer.

Ambient temperature. The surrounding temperature relative to the given point of application.

Antidioxidant. An organic substance that inhibits or retards oxidation.

Blemish. A mark or deformity that impairs appearance.

Blister. A raised spot in the surface or a separation between layers, usually forming a void or air-filled space in the vulcanized article.

Bond. The attachment of a given elastomer to some other member.

Brittleness. The tendency to crack when deformed.

Buna-N. See *Nitrile*.

Buna-S. A general term for the copolymers of butadiene and styrene, also known as *SBR* and *GRS*.

Butaprene. See *Nitrile*.

Butyl. A copolymer of iso-butylene and isoprene.

Cold resistance. Able to withstand the effects of cold or low temperatures without loss of serviceability.

Compression set. The amount by which a rubber specimen fails to return to its original shape after release of compressive load.

Conductive rubber. A rubber capable of conducting electricity. Most generally applied to rubber products used to conduct static electricity.

Copolymer. A polymer consisting of two different monomers chemically combined.

Creep. The progressive relaxation of a given rubber material while under stress. This relaxation eventually results in permanent deformation, or "set."

Cross-linking agents. A chemical, or chemicals, that bonds the polymer chains together to form a thermoset rubber product.

Cure. See *Vulcanization.*

Cure date. The date when an O-ring was molded; for example, "4 Q 96" means "fourth quarter, 1996."

Curing temperature. The temperature at which the rubber product is vulcanized.

Damping. The quality of an elastomer to absorb forced vibrational energy.

Degassing. The intentional, but controlled, outgassing of a rubber substance or other material.

Dielectric properties. The ability of a material to resist puncture due to electric stress.

Diffusion. The mixing of two or more substances (solids, liquids, gasses, or combinations of these) due to the intermingling motion of their individual molecules. Gasses diffuse more readily than solids.

Durometer. An instrument for measuring the hardness of a rubber; measures the resistance to the penetration of an indenter point into the surface of the rubber; the numerical scale of rubber hardness.

Dynamic. An application in which the seal is subject to movement or moving parts contact the seal.

Dynamic packing. A package employed in a joint whose members are in relative motion.

Dynamic seal. A seal required to prevent leakage past parts in relative motion.

Elasticity. The tendency of an article to return to its original shape after deformation.

Elastomer. Any natural or synthetic material with resilience or memory sufficient to return to its original shape after major or minor distortion.

Elongation. Generally, *ultimate elongation*, or the percent of increase in the original length of a specimen when it breaks.

EPDM (EPT, Nordel DuPont Co.). Terpolymer of ethylene-propylene-diene (noted for excellent ozone resistance).

Evaporation. The direct conversion from the liquid to vapor state of a given fluid.

Extrusion. Distortion or flow, under pressure, of a portion of a seal into clearance between mating parts.

Feather edge. The sharp, thin edge on parts, such as wiper seals and cups, also called "knife edge."

Flame resistance. The resistance to burning of material that will not withstand combustion under ordinary conditions.

Flash. Excess rubber left around a rubber part after molding, due to space between mating mold surfaces, removed by trimming.

Flex cracking. A surface cracking induced by repeated bending or flexing.

Flex resistance. The relative ability of a rubber article to withstand dynamic bending stress.

Flow. Ability of heated plastic or uncured rubber to travel in the mold and runner system during the molding process.

Flow cracks. Surface imperfections due to improper flow and failure of stock to knit or blend with itself during the molding operation.

Fluid. A liquid or a gas.

Fluorocarbon (Viton DuPont Dow Elastomers, Fluorel 3 M Co.). A polymer designed to meet the most rigid requirements for use in oils, solvents, synthetic lubricants, and corrosive chemicals, at elevated temperatures.

Friction. Resistance to motion due to contact of surfaces.

Friction, break out. Friction developed during the initial or starting motion.

Friction, running. Constant friction developed during operation of a dynamic O-ring.

Fuel, aromatic. Fuel that contains benzene or aromatic hydrocarbons; causes little swell of rubber.

Fuel, nonaromatic. Fuel composed of straight-chain hydrocarbons; causes little swell of rubber.

Gasket. A device used to retain fluids under pressure or seal out foreign matter. Normally refers to a static seal.

Gas permeability. The degree to which a substance resists permeation of gas under pressure.

Gland. The cavity into which an O-ring is installed, includes the groove and mating surface of the second part, which together confine the O-ring.

GRS. See *SBR.*

Hardness. Resistance to a disturbing force. Measured by the relative resistance of a material to an intender point of any of a number of standard hardness testing instruments, see *Durometer.*

Hardness, Shore A. The rubber durometer hardness as measured on a Shore Wilson-Shore Instruments A Gauge. Higher numbers indicate harder materials; lower numbers, softer materials.

Heat aging. A test for degradation of physical properties as a result of exposure to high-temperature conditions.

Heat deflection temperature. The temperature at which a standard plastic test bar deflects 0.010 in. under a stated load of either 66 psi or 264 psi.

Hermetic seal. An airtight seal having no detectable leakage.

Homogeneous. In general, a material of uniform composition throughout; in seals, a rubber seal without fabric or metal reinforcement.

Hydrocarbon solvents, aromatic. Solvents having a basic benzene structure, usually coat tar types such as benzene or toluene orxylene.

Hypalon. DuPont trade name for chlorosulfonated polyethylene, an elastomer.

Identification. Colored dots or stripes on seals for identification purposes, seldom used.

Immediate set. The deformation found by measurement immediately after removal of the load causing the deformation.

Immersion. Placing an article into fluid, generally so it is completely covered.

Impact. The single, instantaneous stroke or contact of a moving body with another, either moving or at rest, such as a large lump of material dropping on a conveyor belt.

IRHD (international rubber hardness degree). A method to characterize an elastomer by its resistance to penetration of a known geometry indenter by a known force. The micro technique is reproduced on irregular, as well as flat, surfaces, and on cross-sections as small as 1 mm in thickness (04 in.). The readings are similar, but not identical to Shore A, see *Durometer.*

Injection molding. Molding in which the rubber or plastic stock is heated and, while in the flowable state, forced or injected into the mold cavity.

Insert. Typically, a metal or plastic component to which rubber or plastic is chemically or physically bonded during the molding process.

Leakage rate. The rate at which a fluid (either gas or liquid) passes a barrier. The total leakage rate includes the amounts that diffuse or permeate the material of the barrier as well as the amount that escapes around it.

Life test. A laboratory procedure used to determine the amount and duration of resistance of an article to specific sets of destructive forces or conditions.

Linear expansion. Expansion in any one linear dimension or the average of all linear dimensions.

Low-temperature flexibility. The ability of a rubber product to be flexed, bent, or bowed at low temperature without cracking.

Mechanical bond. A method of physically bonding rubber to inserts through holes, depressions, or projections in the insert.

Memory. The tendency of a material to return to original shape after deformation.

Microhardness. An electronic measurement of rubber hardness for specimens below 0.25 in. in thickness. Micro hardness, like Shore A and durometer tests, also measures indentation.

Modulus. Tensile stress at specific elongation, usually 100% elongation for elastomers.

Modulus of elasticity. One of several measurements of stiffness or resistance to deformation but often incorrectly used to indicate specifically static tension modulus.

Mold cavity. Hollow space, or cavity, in the mold used to impart the desired form to the product being molded.

Mold finish. The uninterrupted surface produced by intimate contact of rubber with mold surface at vulcanization.

Mold lubricant. A material usually sprayed onto the mold cavity surface prior to the introduction of the uncured rubber to facilitate the easy removal of the molded part.

Neoprene DuPont (GR-M). A polymer of chloroprene prepared from coal, salt, and limestone.

Nitrile. The most commonly used elastomer for O-rings because of its resistance to petroleum fluids, its good physical properties, and its useful temperature range; see also *Buna-N* and *Butaprene.*

Oil resistant. Ability to vulcanize rubber to resist the swelling and the deteriorating effects of various types of oils.

Oil swell. The change in volume of a rubber article due to absorption of oil or other fluid.

Optimum cure. State of vulcanization at which the most desirable combination of properties is attained.

O-ring. A torus; a circle of material with a round cross-section that effects a seal through squeeze or pressure.

O-ring seal. The combination of a gland and O-ring providing a fluid tight closure. (Some designs permit minimum leakage.)

O-ring seal, moving (dynamic). O-ring seal in which there is relative motion between some gland parts and the O-ring; oscillating, reciprocating, or rotary motion.

O-ring seal, nonmoving (static). O-ring seal in which there is no relative motion between any part of the gland and the O-ring. (Distortion from fluid pressure or swell from fluid immersion is excluded.)

Outgassing. A vacuum phenomenon wherein a substance spontaneously releases volatile constituents in the form of vapors or gases. In rubber compounds, these constituents may include water vapor, plasticizers, air, inhibitors, and the like.

Oxidation. The reaction of oxygen on a compound, usually detected by a change in the appearance or feel of the surface, by a change in the physical properties, or both.

Ozone resistance. Ability to withstand the deteriorating effect of ozone, which generally causes cracking.

Packing. A flexible device used to retain fluid under pressure or seal out foreign matter; normally refers to a *dynamic seal.*

Permanent set. The deformation remaining after a specimen has been stressed in tension for a definite period and released for a definite period.

Permeability. The rate at which liquid or gas, under pressure, passes through a solid material by diffusion and solution. In rubber terminology, the rate of gas flow, expressed in atmospheric cubic centimeters per second, through an elastomeric material, 1 cm^2 and 1 cm thick.

pH. Specification of the concentration of either an acid or a base.

Pit (or pock) **mark.** A circular depression, usually small.

Plasticity. When subject to sufficient shearing stress, any given body deforms. After the stress is removed, if there is no recovery, the body is completely plastic. If recovery is complete and instantaneous, the body is completely elastic. A balance between the two is required.

Plasticizer. A substance, usually a heavy liquid, added to an elastomer to decrease stiffness, improve low-temperature properties, or improve processing.

Polymer. A material formed by joining together many (poly) individual units (mer) of one or more monomers, synonymous with *elastomer*.

Polymerization. Chemical reaction whereby either one or more simple materials are converted to a complex material that possesses properties entirely different from the original materials used to start the reaction.

Polyurethane. An organic material noted for its high abrasion, ozone, corona, and radiation characteristics.

Porosity. The quality or state of being porous.

Postcure. The second step in the vulcanization process for some specialized elastomers; provides stabilization of parts and drives off decomposition products resulting from the vulcanization process.

Pure gum state. A nonpigmented, translucent basic polymer. In elastomers, independent of type of radiation specimen, denoting an energy absorption level of 100 ergs per gram of elastomer, approximately equal to 1.2 roentgens.

Rebound. A measure of the resilience, usually as a percentage of vertical return of a body that has fallen and bounced.

Reinforcement agent. Material dispersed in an elastomer to improve compression, shear, or other stress properties.

Relative humidity. The ratio of the quantity of water vapor actually present in the atmosphere to the greatest amount possible at a given temperature.

Resilience. The ability of an elastomer to return to its original size and shape after deforming forces are removed; generally expressed in percent of the ratio of energy removed to the energy used in compressing. ("Resilient" means having that capability.)

Rotary seal. A seal, such as an O-ring or a quad-ring seal, exposed on either the ID or OD sealing surface to a rotating component, such as shaft seals.

Rough trim. Removal of superfluous material by pulling or picking; usually the removal of a small portion of the flash or sprue that remains attached to the product.

Rubber. See *Elastomer*.

Rubber, natural. Raw or crude rubber obtained from vegetable sources.

Rubber, synthetic. Manufactured or human-made elastomers.

SBR. Copolymer of butadiene and styrene; an all-purpose synthetic, similar to natural rubber. (Butadiene is a gaseous material of petroleum; styrene, a reaction product of ethylene and benzene.)

Scorching. Premature curing or setting up of a raw compound during processing.

Seal. Any device used to prevent the passage of a fluid, gas, or liquid.

Service. Operating conditions to be met.

Shaft. Reciprocating or rotating member, usually within a cylinder; not in direct contact with the walls.

Shelf aging. The change in a material's properties that occurs in storage with time.

Shrinkage. The ratio between a mold cavity size and the size of a product molded in that cavity; decreased volume of a seal, usually caused by extraction of soluble constituents by fluids followed by air drying.

Silicone rubber. Elastomer that retains good properties through extra wide temperature ranges.

Sorption. A term used to denote the combination of adsorption and absorption processes in the same substance.

Specific gravity. The ratio of the weight of a given substance to the weight of an equal volume of water at any specific temperature. Sphericity is the measure of a tolerance of a molded ball, or ground ball, in reference to a perfect sphere; also described as "roundness."

Spiral twist. A type of seal failure in reciprocating applications that results from the twisting action that strains or ruptures the rubber.

Squeeze. Cross section diametrical compression of an O-ring between the surface of the groove bottom and the surface of the other mating metal part in the gland assembly.

Static seal. The part designed to seal between parts having relatively no motion.

Strain. Deflection due to force.

Stress. Force per unit of original cross-section area.

Stress relaxation. Decreasing stress with constant strain over a given time interval; a viscoelastic response.

Sun checking. Surface cracks, checks, or grazing caused by exposure to direct or indirect sunlight.

Surface finish. A numerically averaged value of surface roughness, generally in units of microinches or micrometers.

Swell. Increased volume of a specimen, caused by immersion in a fluid (usually liquid).

Tack. The degree of adhesion of materials of identical nature to each other.

Tear resistance. Resistance to growth of a cut or nick when tension is applied to the cut specimen; commonly expressed as pounds per square inch of thickness.

Tear strength. The force required to rupture a sample of stated geometry.

Temperature range. Maximum and minimum temperature limits in which a seal compound functions in a given application.

Tensile strength. Force, in pounds per square inch, required to cause the rupture of a specimen of rubber material.

Tension modulus. Resistance to being stressed; defined as the force, in pounds, necessary to stretch a piece of rubber, 1 in.^{2} in cross-section, a specified amount. Normally expressed as a percentage of original length, and the stress as pounds per square inch at the fixed elongation.

Terpolymer. A polymer consisting of three different monomers chemically combined.

Thermal expansion. Expansion caused by an increase in temperature, may be linear or volumetric.

Thermoplastic. A plastic capable of being repeatedly softened by increase in temperature and hardened by decrease in temperature.

Thermoplastic rubber. Rubber that does not require chemical vulcanization and repeatedly softens when heated and stiffens when cooled, exhibiting only slight loss of original characteristics.

Thermoset. An elastomer or plastic cured under application of heat or chemical means to make a product substantially infusible or insoluble.

Threshold. The maximum tolerance of an elastomer to radiation dosage, expressed as a total number of ergs per gram (or rads) beyond which physical properties are significantly degraded. This is generally an arbitrary value, depending on the function and environment.

Torsion strength. The ability of rubber to withstand twisting.

Transfer molding. A method of molding in which material is placed in a pot, located between the top plate and plunger, and squeezed from the pot through gates (or sprues) into the mold cavity.

Trapped air. Air trapped in a product or a mold during cure. Usually causing a loose ply or cover, or a surface mark, depression, or void.

Trim. The process involving removal of mold flash.

Trim out. Damage to mold skin or finish by close trimming.

Ultimate elongation. A measure of how far a material stretches before breaking; expressed as a percentage of its original length.

Undercure. Degree of a cure less than optimal; may be evidenced by tackiness, loginess, or inferior physical properties.

Vacuum. A defined space occupied by a gas at less then atmospheric pressure.

Vapor pressure. The maximum pressure exerted by a liquid or a solid heated to a given temperature in a closed container.

Vibration dampening. The ability of an elastomer to absorb vibrations or shocks.

Viscosity. The property of fluids and plastic solids by which they resist an instantaneous change of shape, that is, resistance to flow.

Void. The absence of material or an area devoid of materials where not intended.

Volatilization. The transition of either a liquid or solid directly into a vapor state. In the case of a liquid, this transition is called *evaporation*, whereas in the case of a solid, it is termed "sublimation."

Volume change. A change in the volume of a seal as result of its immersion in a fluid; expressed as a percentage of the original volume.

Volume swell. An increase in the physical size caused by the swelling action of a liquid.

Vulcanization. A thermosetting reaction involving the use of heat and pressure, resulting in greatly increased strength and elasticity of rubberlike materials.

Vulcanizing agent. A material that causes the vulcanization of an elastomer.

Weathering. The detrimental effect on an elastomer or plastic after outdoor exposure.

Weather resistance. The ability to withstand weathering factors, such as oxygen, ozone, atmospheric pollutants, erosion, temperature cycling, and ultraviolet radiation.

Wiper ring. A ring employed to remove excess fluid, mud, and the like from a reciprocating member before it reaches the packings.

6. ABBREVIATIONS

These are the abbreviations commonly used to describe nonmetallic materials:

ABR	Acrylate butadiene rubber
ABS	Acrylonitrile butadiene styrene

ASA	Acrylonitrile styrene acrylate
(A)U, (E)U	Polyurethane AU (polyester), EU (polyether)
BR	Butadiene rubber
BS	Butadiene styrene
CA	Cellulose acetate
CAB	Cellulose acetate butyrate
CAP	Cellulose acetate propionate
CFM	Polychlorotrifluoroethylene
CM	Chloropolyethylene
CP	Cellulosepropionate
CPE	Chlorinated polyethylene
CPVC	Chlorinated polyvinylchloride
CR	Chloroprene rubber
CSM	Chlorosulphonated polyethylene
DAP	Diallyl phthalate
ECTFE	Ethylenechlorotrifluoroethylene
EPDM	Ethylene propylene rubber
EPR	Ethylene propylene rubber
EPS	Expanded polystyrene
ESC	Environmental stress cracking
ETFE	Ethylene tetrafluoroethylene
EVA	Ethylene vinylacetate
EVAC	Ethylene vinylacetate
EVAL	Ethylene vinylalcohol
FEP	Fluorinated ethylene propylene
FFKM	Perfluoro elastomer
FKM	Fluorocarbon copolymer
FMK	Fluor-silicone rubber
FPA	Perfluoralkoxy
FRP	Fiber-reinforced plastic
GR-A	Apolybutadiene acrylonitrile rubber
GRE	Glass-reinforced epoxy
GR-I	Butyl rubber, polyisobutylene isoprene rubber
GR-N	Nitrile rubber, nitrile butadiene rubber, polybutadiene acrylonitrile rubber
GRP	Glass-reinforced plastic
GR-S	Styrene butadiene rubber, polybutadiene styrene rubber
GRUP	Glass-reinforced unsaturated polyester
GRVE	Glass-reinforced vinyl ester

HDPE	High-density polyethylene
IIR	Butyl rubber
IM	Polyisobutene rubber
IR	Isoprene rubber
ISO	International Standards Organization
MDPE	Medium-density polyethylene
MF	Melamine formaldehyde
NBR	Nitrile butadiene rubber
NR	Natural rubber
PA	Polyamide
PAI	Polyaramide imide
PAN	Polyacrylonitrile
PB	Polybutylene
PBTP	Polybutylene terephthalate
PC	Polycarbonate
PCTFE	Polychlorotrifluoroethylene
PEEK	Polyetheretherketone
PEI	Polyetherimide
PES	Polyethersulfone
PETP	Polyethylene terephthalate
PEX	Cross-linked polyethylene
PF	Phenol formaldehyde
PFA	Perfluoroalkoxy copolymer
PFEP	Fluorinated ethylene propylene
PI	Polyimide
PIB	Polyisobutylene
PIR	Poly-isocyanurate rubber
PK	Polyketone
PMMA	Polymethyl methacrylate
POM	Polyoxymethylene, polyformaldehyde
PP	Polypropylene
PPO	Polyphenylene oxide
PPS	Polyphenelynesulphone
PS	Polystyrene
PSU	Polysulfone
PTFE	Polytetrafluoridethylene
PUF	Polyurethane (foam)
PUR	Polyurethane
PVAC	Polyvinyl acetate
PVAL	Polyvinyl alcohol

PVC	Polyvinylchloride
PVCC	Chlorinated polyvinyl chloride
PVDC	Polyvinylidene chloride
PVDF	Polyvinylidenefluoride
PVF	Polyvinyl fluoride
SAN	Styrene acrylonitrile
SB	Styrene butadiene
SBR	Styrene butadiene rubber
SI	Silicone
SIC	Silicon carbide
TFE	Polytetrafluoroethylene
TPE	Thermoplastic elastomers
TPU	Thermoplastic polyurethane
UF	Ureum formaldehyde
UHMWHDPE	Ultra-high molecular weight, high-density polyethylene
UP	Unsaturated polyester
UPVC	Unplasticised polyvinylchloride
UV	Ultraviolet light
VAC	Vinylacetate
VC	Vinylchloride
XLPE, PEX	Cross-linked polyethylene consisting of long polymer chains in a 3-dimensional structure
XPS	Extruded polystyrene

7. LIST OF COMMERCIALLY AVAILABLE NONMETALLIC MATERIALS

Trade Name	Chemical Classification	Manufacturer
Acalor	Resin-filled cement	Acalor, England
Adiprene	Polyurethane rubber	DuPont, USA
Aerophenal	Polyfluoride	Ciba-Geigy
Akulon	Polyamide	AKZO, Netherlands
Alathon	Polyethylene	DuPont, USA
Albertol	Saturated polyesters	Hoechst, Germany
Algoflon	Polytetrafluoroethylene	Montedison, Italy
Alkathene	LD polyethylene	ICI
Alkon	POM	ICI
Alnovol	Phenolics	Hoechst, Germany

Alpolit	Unsaturated polyesters	Hoechst, Germany
Alresen	Phenolic, modified	Hoechst, Germany
Altuglas	Polymethyl metacrylate	Elf Atochem, France
Amilan	Polyamide	Toray Industries, Japan
Ampal	Unsaturated polyesters	Ciba-Geigy, Switzerland
Ampcoflex	Polyvinyl chloride	Atlas Plastics, USA
Appryl	Polypropylene	Atochem
Araldit	Epoxies	Ciba-Geigy, Switzerland
Ardel	Polyarylate	Amoco, USA
Arenka	Polyamide	AKZO, Netherlands
Arnite	Unsaturated polyesters	AKZO, Netherlands
Arnitel	Saturated polyester	AKZO, Netherlands
Arylon	Polyarylether, polyarylates	DuPont, USA
Asplit	Resin filled cement	Hoechst, Germany
Astraglas	Polyvinyl chloride (soft)	Dynamit Nobel
Astralit	Polyvinyl chloride (hard)	Dynamit Nobel
Astralon	Polyvinyl chloride	Hüls, Germany
Astratherm	Polyvinyl Chloride (hard)	Dynamit Nobel
Atlac	Unsaturated polyesters	DSM, Netherlands
Bakelite	Phenolics	Bakelite, Germany
Basopor	UF	BASF
Basotect	UF	BASF
Bayblendt	PC/ABS blend	Bayer
Baydur	Polyurethanes	Bayer, Germany
Bayflex	Polyurethanes	Bayer, Germany
Baygal	PUR	Bayer
Baylon	HDPE	Bayer
Baymer	Polyisocyanurate	Bayer, Germany
Baymidur	PUR	Bayer, Germany
Baypren	Polychloroprene	Bayer, Germany
Baysilone	Silicones	Bayer, Germany
Beckocoat	Polyurethanes	Hoechst, Germany
Beckopox	Epoxies	Hoechst, Germany
Beckurol	Ureas	Hoechst, Germany
Beetle	Unsaturated polyesters, phenolics	BP Chemicals, England
Benvic	Polyvinylchloride	Solvay, Belgium
Bondstrand	Fiber-reinforced plastic piping	Ameron, USA
Bornum harz	Resin impregnated graphite	HarzerAchsenwerke, Germany
Breon	Polybutadiene acrylonitrile	Zeon, Germany
Budene	Polybutadiene	Goodyear, USA

(Continues)

(Continued)

Trade Name	Chemical Classification	Manufacturer
Buna	Polybutadiene	Hüls, Germany
Calibre	PC	DOW
Capron	Polyurethanes	Allied Corp., USA
Caradate	Isocyanates for polyurethanes	Shell
Caradol	Polyols for polyurethanes	Shell
Carbofrax	Silicon carbide	Carborundum, USA
Cariflex	Polybutadiene, stryrene elastomers	Shell
Carilon	Polyketone	Shell
Carina	Polyvinyl chloride	Shell
Carinex	Polystyrene	Shell
Carlona	Polyethylene	Shell
Carlona P	Polypropylene	Shell
Casocryl	Polymethyl methacrylate	Elf Atochem, France
Celcon	Polyformaldehyde	Hoechst, Germany
Cellasto	PUR	BASF
Cellidor b	Cellulose acetate butyrate	Albis Plastics, Germany
Cibamin	Ureas, melamines	Ciba-Geigy, Switzerland
Cibanoid	UF	Ciba-Geigy
Conapoxy	Melamines	Conap, USA
Coroplast	Polyvinylchloride	Coroplast, Germany
Corvic	Polyvinylchloride	ICI, England
Courtelle	Polyacrylonitrile	Courtaulds, England
Crastin	PET/PBT	Ciba-Geigy
Crylor	Polyacrylonitrile	Rhone Poulenc, France
Crystic	Unsaturated polyesters	Scott Bader Co., England
Cycolac	Acrylonitrile butadiene styrene	General Electric, USA
Dacron	Saturated polyesters	DuPont, USA
Daplen	Polypropylene	PCD Linz, Austria
Darvic	Polyvinylchloride	Weston Hyde, England
Degalan	Polymethyl methacrylate	Degussa, Germany
Delpet	Polymethyl methacrylate	Asahi Chem., Japan
Delrin	Polyformaldehyde	DuPont, USA
Derakene	Unsaturated polyesters, vinylester type	DOW, USA
Desmodur	Isocyanates for polyurethanes	Bayer, Germany
Desmopan	Polyurethane rubber	Bayer, Germany
Desmophen	Polyols for polyurethanes	Bayer, Germany
Dewoglas	Polymethyl methacrylate	Degussa, Germany
Diabon	Graphite	Sigri, Germany

Diakon	Polymethyl methacrylate	ICI, England
Dobeckan	Unsaturated polyesters, polyurethanes	BASF, Germany
Dolan	Polyacrylonitrile	Hoechst, Germany
Dorix	Polyamide	Bayer, Germany
Dorlastan	Polyurethane rubber	Bayer, Germany
Dowlex	PE	DOW
Dpc 2000 T	LDPE foil	ICl
Drakaflex	Polyurethanes	Draka, Netherlands
Dralon	Polyacrylonitrile	Bayer, Germany
Durabon	Carbon	Sigri, Germany
Duran 50	Glass	Jena Glaswerk Schott, Germany
Durel	Polyarylate	Hoechst, Germany
Durethan	Polyamide	Bayer, Germany
Durolon	PC	Montedison
Durophen	Phenolics	Hoechst, Germany
Dutral	EP	Montedison
Dyflor	PVDF	Dynamit Nobel
Dylene	Polystyrene, styrene acrylonitrile	ARCO Polymers, USA
Dynapol	Saturated polyesters	Hüls, Germany
Edifran	PCTFE	Montedison
Edistir	Polystyrene	Enichem, Italy
Editer	ABS	Montedison
Ekavyl	Polyvinylchloride	Elf Atochem, France
Elastan	PUR	BASF
Elastocoat	PUR	BASF
Elastoflex	PUR	BASF
Elastofoam	PUR	BASF
Elastogran	PUR	BASF
Elastolit	PUR	BASF
Elastollan	Polyurethanes	Elastogran, Germany
Elastopal	PUR	BASF
Elastopan	PUR	BASF
Elastopor	PUR	BASF
Elastosil	Silicone rubber	Wacker-Chemie, Germany
Elasturan	PUR	BASF
Elexar	Styrene butadiene, styrene rubber	Shell
Eltex	Polyethylene	Solvay, Belgium
Eltex p	Polypropylene	Solvay, Belgium
Elvanol	Polyvinylalcohol	DuPont, USA
Epikote	Epoxies	Shell
Epon	Epoxies - USA	Shell

(Continues)

(Continued)

Trade Name	Chemical Classification	Manufacturer
Eraclear	LDPE	Enichem
Eraclene H	HDPE	Enichem
Eriflon PVDF	PVDF	Solvay
Ertalon	PA	AKZO
Ertalon	PA	Atochem
Ertalon	PA	BASF
Ertalon	PA	DSM
Escorene	Polyethylene	Exxon, USA
Extir	EPS	Montedison
Fertene	LDPE	Montedison
Fibercast	Fiber-reinforced epoxies	Fibercast, USA, Germany
Finathene	Polyethylene	Fina, Belgium
Fluon	Polytetrafluoroethylene	ICI, England
Fluorel	Vinylide fluoride-hexafluoropropylene	3M, USA
Fluoroflex	Fluorinated polymers	Resistoflex, USA, Germany
Fluorogreen	Fluorinated polymers	Peabode Dore, USA
Fluoroline	Fluorinated polymers	BTR, England
Fluorosint	Fluorinated polymers	Polypenco, Germany
Foraflon	Polyvinylidene fluoride	Elf Atochem, France
Formica	Melamines	Formica Corp., USA
Fortiflex	HDPE	Solvay
Fortilene	PP	Solvay
Fortron	PPS	Hoechst
Furacin	Furane-filled cement	Prodorite, England
Gabrite	UF	Montedison
Gaflon	Polytetrafluoroethylene	Plastic Omnium, France
Gemon	Polyimide	General Electric, USA
Geon	Polyvinylchloride	B.F. Goodrich, USA
Glad	Polyethylene	Union Carbide, USA
Goretex	Polytetrafluoroethylene	W.L. Gore, USA
Granlar	LCP	Montedison
Graphilor	Resin-impregnated graphite	LeCarbone-Lorraine, France
Grilamid	Polyamide	EMS-Chemie, Switzerland
Grillodur	Unsaturated polyesters	Grillo-Werke, Germany
Halar	Polytrifluoroethylene	Ausimont, USA
Halon	Polytetrafluoroethylene	Ausimont, USA
Haveg	Phenolics, furanes	Haveg, USA
Herox	Polyamide	DuPont, USA
H.E.T.	Chlorinated unsaturated terpolymer	Ashland Chem., USA
Hetron	Chlorinated unsaturated polyesters	Ashland Chem., USA

Hfr cement	Potassium silicate cement	Hoechst, Germany
Hostadur	PBT, PET	Hoechst
Hostaflex	Polyvinylchloride	Hoechst, Germany
Hostaflon	Polytetrafluoroethylene	Hoechst, Germany
Hostaflon-c	Polychlorotrifluoroethylene	Hoechst, Germany
Hostaform	POM	Hoechst
Hostalen	Polyethylene	Hoechst, Germany
Hostalen gur	UHMW PE	Hoechst
Hostalen LD	LDPE	Hoechst
Hostalen-PP	Polypropylene	Hoechst, Germany
Hostalit	Polyvinylchloride	Hoechst, Germany
Hostapor	EPS	Hoechst
Hostapox	EP	Hoechst
Hostapren	CPE	Hoechst
Hostaset PF	PF	Hoechst
Hostaset UF	UF	Hoechst
Hostaset UP	UP	Hoechst
Hostatec	PEK	Hoechst
Hostyren	Polystyrene	Hoechst, Germany
Hostyren XS	SB	Hoechst
Hycar	Polybutadiene, stryrene elastomers	B.F. Goodrich, USA
Hypalon	Chlorosulphonated polyethylene	DuPont, USA
Hytrel	Saturated polyesters	DuPont, USA
Hyvis	Polyisobutylene	BP Chem., England
Icdal	Polyimide	Hüls, Germany
Imipex	Polyimide	General Electric, USA
Impet	PET	Hoechst
Impolex	Unsaturated polyesters	ICI, England
Inklurit	UF	BASF
Ixan	Polyvinylidene chloride	Solvay, Belgium
Kalrez	Perfluoro elastomer	DuPont, USA
Kamax	Polyimide	Rohm and Haas, USA
Kapton	Polyimide	DuPont, USA
Karbate	Resin-impregnated graphite	Union Carbide, USA
Keebush	Resin-impregnated graphite	APV-Kester, England
Kel-F	Polychlorotrifluoroethylene	3M, USA
Keltan	Ethylene propylene diene terpolymer	DSM, Netherlands
Kematal	POM	Hoechst
Keranol	Resin-filled cement	Keramchemie, Germany
Kerimid	Polyimide	Rhone-Poulenc, France
Kermel	Polyimide	Rhone-Poulenc, France
Kevlar	Polyaramide (fiber)	DuPont, USA

(Continues)

(Continued)

Trade Name	Chemical Classification	Manufacturer
Kinel	Polyimide	Rhone-Poulenc, France
Kobiend	PC/ABS blend	Montedison
Kralastic	Acrylonitrile butadiene styrene	Uniroyal, Japan
Kraton g	Styrene butadiene styrene rubber	Shell
Kydex	Polyvinylchloride	Rohm and Haas, USA
Kynar	Polyvinylidene fluoride	Elf Atochem, France
Lacqrene	PS	Atochem
Lacqtene	Polyethylene	Elf Atochem, France
Lacqvyl	PVC	Atochem
Lamellon	Unsaturated polyesters	—
Larflex	EP	Lati
Laril	PPO	Lati
Laroflex	Polyvinylchloride	BASF, Germany
Larton	PPS	Lati
Lastane	PUR	Lati
Lastiflex	ABS/PVC blend	Lati
La Stil	SAN	Lati
Lastilac	ABS	Lati
Lastilac 10	ABS/PC blend	Lati
Lastirol	PS	Lati
Lasulf	PSU	Lati
Latamid	PA	Lati
Latan	POM	Lati
Latene	PP	Lati
Latene HD	HDPE	Lati
Later	PBT	Lati
Latilon	PC	Lati
Leacril	Polyacrylonitrile	—
Legupren	Unsaturated polyesters	Bayer, Germany
Leguval	Unsaturated polyesters	DSM, Netherlands
Lekutherm	Epoxies	Bayer, Germany
Levaflex	TPO	Bayer
Levepox	Epoxies	Bayer, Germany
Lexan	Polycarbonate	General Electric, USA
Lexgard	PC	GEP
Linatex	Natural rubber, soft	WilkinsonRubberLinatex
Lucalor	CPVC	Atochem
Lucite	Polymethyl methacrylate	DuPont, USA
Lucolene	PVC (soft)	Atochem
Lucorex	Polyvinylchloride	Elf Atochem, France
Lucovyl	PVC	Atochem
Lucovyl	PVC	Rhone-Poulenc

Lupolen	Polyethylene	BASF, Germany
Luran	Styrene acrylonitrile	BASF, Germany
Luranyl	PPE	BASF
Lustran	Styrene acrylonitrile	Monsanto, USA
Lustrex	Polystyrene	Monsanto, USA
Luxor	PS, SAN	Montedison
Lycra	Polyurethanes	DuPont, USA
Madurit	Melamines	Hoechst, Germany
Magnum	ABS	Dow, Netherlands
Makroblend	PC blend	Bayer
Makrofol	PC foil	Bayer
Makrolon	Polycarbonate	Bayer, Germany
Manolene	PE	Rhone-Poulenc
Maprenal	Melamines	Hoechst, Germany
Maranyl	Polyamides	ICI, England
Melaplast	MF	Bayer
Melbrite	Melamines	Montedison, Italy
Melinex	Saturated polyesters	ICI, England
Melmex	Melamines	BP Chemicals, England
Melopas	Melamines	Ciba-Geigy, Switzerland
Menzolit	Epoxies, unsaturated polyesters	Menzolit-Werke, Germany
Minlon	Polyamides	DuPont, USA
Mipolam	Polyvinylchloride	Hüls, Germany
Mipoplast	PVC soft	Dynamit Nobel
Moltapren	Polyurethane foam	Bayer, Germany
Moltopren	PUR	Bayer
Moplen	Polypropylene	Himont, Italy
Mouldrite	UF	ICI
Mowilith	Polyvinylacetate	Hoechst, Germany
Mowiol	Polyvinylalcohol	Hoechst, Germany
Mylar	Saturated polyesters	DuPont, USA
Nandel	Polyacrylonitrile	DuPont, USA
Napryl	Polypropylene	Elf Atochem, France
Natene	Polyethylene	Elf Atochem, France
Natsyn	Polyisoprene	Goodyear, USA
Neonit	EP	Ciba-Geigy
Neopolen	PE foam	BASF
Neoprene	Polychloroprene	DuPont, USA
Nitril	Polybutadiene acrylonitrile	—
Nivionplast	PA	Enichem
Nordel	Ethylene-propylene diene terpolymer	DuPont, USA
Noryl	Polyphenylene oxide	General Electric, USA
Novodur	Acrylonitrile butadiene styrene	Bayer, Germany

(Continues)

(Continued)

Trade Name	Chemical Classification	Manufacturer
Novolen	Polypropylene	BASF, Germany
Novolux	Polyvinylchloride	Weston Hyde, England
Nylon	Polyamide	DuPont, USA
Nyrim	Polyamide	DSM, Netherlands
Oppanol	Polyisobutylene	BASF, Germany
Orbitex	Epoxies	Ciba-Geigy, Switzerland
Orgalloy	PA/PP blend	Atochem lend
Orgamide	PA	Atochem
Orgasol	PE or coPA	Atochem
Orgater	Polycarbonate	Elf Atochem, France
Orgavyl	Polyvinylchloride	Elf Atochem, France
Orlon	Polyacrylonitrile	DuPont, USA
Oroglas	Polymethyl methacrylate	Rohm and Haas, USA
Palapreg	UP	BASF
Palatal	Unsaturated polyesters	BASF, Germany
Pan	Polyacrylonitrile	Bayer, Germany
Paraplex	Unsaturated polyesters	Rohm and Haas, USA
Parylene	Polyarylene	Union Carbide, USA
Peek	Polyetheretherketone	ICI, England
Pellethane	TPU	DOW
Penton	Polydichloromethyloxetane	—
Perbunan	Polybutadiene acrylonitrile	Bayer, Germany
Perlon	Polyamide	Perlon, Germany
Perspex	Polymethyl methacrylate	ICI, England
Petion	PET	Bayer
Pibiter	PBT	Montedison
Plaskon	Ureas	Plaskon, USA
Plastopal	Ureas	BASF, Germany
Plexidur	Polymethyl methacrylate	Rohm and Haas, USA
Plexiglas	Polymethyl methacrylate	Rohm and Haas, USA
Plioflex	Polybutadiene styrene	Goodyear, USA
Pocan	Saturated polyesters	Bayer, Germany
Pollopas	UF	Dynamit Nobel
Polydur	Unsaturated polyesters	Hüls, Germany
Polylite	Unsaturated polyesters	Reichhold Chem., USA
Polystyrol	Polystyrene	BASF, Germany
Polyviol	Polyvinyl alcohol	Wacker-Chemie, Germany
Primef	PPS	Solvay
Propathene	Polypropylene	ICI, England
Puise	PC/ABS blend	Dow
Pyrex	Glass	Sovirel, France
Quacorr	Furanes	PO Chemicals, USA
Quickfit	Glass	Corning, England
Radel	Polyarylether	Amoco, USA

Renolit	Polyvinylchloride	Renolit-Werke, Germany
Renyl	PA6	Montedison
Resamin	Ureas	Hoechst, Germany
Rhenoflex	Polyvinylchloride	Hüls, Germany
Rhepanol	Polyisobutylene sheet	—
Rhodopas	PVC	Rhone-Poulenc
Rhodorsil	Silicone rubbers	Rhone-Poulenc, France
Riblene D	LDPE	Enichem
Rigidex	Polyethylene	BP Chemicals, England
Rilsan	Polyamide	Elf Atochem, France
Ronfalin	ABS	DSM
Rulon	Filled PTFE	Dixon Corp., USA
Rutapox	Epoxies	Bakelite, Germany
Rynite	PBT, PET	DuPont de Nemours
Ryton	Polyphenylene sulphide	Phillips Petr., Belgium
Saran	Polyvinylidene chloride	DOW, USA
Setal	Unsaturated polyesters	Synthese, Netherlands
Setapol	Unsaturated polyesters	Synthese, Netherlands
Shell pb	Polybutene	Shell
Sicron	PVC	Montedison
Silastic	Silicone rubbers	DOW, USA
Silcoset	Silicone rubbers	ICI, England
Silopren	Silicone rubbers	Bayer, Germany
Sinkral	ABS	Enichem
Sinvet	PC	Enichem
Solef	Polyvinylidene fluoride	Solvay, Belgium
Solvic	Polyvinyl chloride	Solvay, Belgium
Soreflon	Polytetrafluoroethylene	Elf Atochem, France
Stamylan	Polyethylene	DSM, Netherlands
Stamylan p	Polypropylene	DSM, Netherlands
Stamylex	LDPE	DSM
Stanyl	Polyamide	DSM, Netherlands
Strातyl	EP	Rhone-Poulenc
Styrocell	Polystyrene foam	Shell
Styrodur	Polystyrene foam	BASF, Germany
Styrofoam	Polystyrene foam	DOW, USA
Styron	Polystyrene	DOW, USA
Styropor	Polystyrene foam	BASF, Germany
Supec	PPS	GEP
SWD cement	Sodium silicate cement	Hoechst, Germany
Synolite	Unsaturated polyesters	DSM, Netherlands
Technyl	Polyamides	Rhone-Poulenc, France
Tedlar	Polyvinylfluoride	DuPont, USA
Tedur	PPS	Bayer
Teflon	Polytetrafluoroethylene	DuPont, USA

(Continues)

(Continued)

Trade Name	Chemical Classification	Manufacturer
Teflon FEP	Fluorinated ethylene propylene	DuPont, USA
Tenax	Carbon fiber	Tenax, Germany
Tenite butyrate	Cellulose acetate butyrate	Eastman Chem. Prod., USA
Tenite CAB	Cellulose acetate butyrate	Eastman Chem. Prod., USA
Tenite PE	Polyethylene	Eastman Chem. Prod., USA
Terblend B	ABS/PC blend	BASF
Terblend S	ABA/PC blend	BASF
Tergal	Saturated polyesters	Rhone-Poulenc, France
Terlenka	Saturated polyesters	ENKA, Germany
Terlenka	PET fiber	AKZO PET
Terluran	Acrylonitrile butadiene styrene	BASF, Germany
Terylene	Saturated polyesters	ICI, England
Ternil	PA6	Montedison
Therban	Polybutadiene acrylonitrile rubber	Bayer, Germany
Thiokol	Polysulphides	Thiokol Corp., USA
Torlon	Polyamide-imide	Amoco Corp., USA
Trevira	Saturated polyesters	Hoechst, Germany
Trocal	Polyvinylchloride	Hüls, Germany
Trocellen	PE foam	Dynamit Nobel
Trogamid	Polyamides	Hüls, Germany
Trolitan	PF	Dynamit Nobel
Trolitul	PS	Dynamit Nobel
Trosiplast	PVC hard	Dynamit Nobel
Trovidur	Polyvinylchloride	Hüls, Germany
Trovidur PP	Polypropylene	Hüls, Germany
Trovipor	PVC foam	Dynamit Nobel
Tufnol	Phenolics, furanes	Tufnol, England
Tufsyn	Polybutadiene	Goodyear, USA
Twaron	Polyaramide (fiber)	AKZO, Netherlands
Tynex	Polyamides	DuPont, USA
Tyril	SAN	Dow
Tyrin	CPE	Dow
Udel	Polysulfone, Polyether sulfone	Amoco, USA
Uformite	Ureas	Reichold, USA
Ugikapon	Unsaturated polyesters	Elf Atochem, France
Ukapor	Polystyrene	Elf Atochem, France
Ultem	Polyetherimide	General Electric, USA

Ultrablend	PBT/PET blend	BASF
Ultrablend S	PBT blend	BASF
Ultradur	Saturated polyesters	BASF, Germany
Ultraform	POM	BASF
Ultramid	Polyamides	BASF, Germany
Ultranyl	PPE/PA blend	BASF
Ultrapas	Melamines	Hüls, Germany
Ultrapek	PEK	BASF
Ultrason E	Polyethersulphone	BASF, Germany
Ultrason S	Polysulphone	BASF, Germany
Ultrax	LCP	BASF
Uralam	Unsaturated polyesters	Synthetic Resins Ltd., England
Ureol	PUR	Ciba-Geigy
Urepan	Polyurethanes	Bayer, Germany
Urtal	ABS	Montedison
Valox	Saturated polyesters	General Electric, USA
Vandar	PBT	Hoechst
Varlan	Polyvinylchloride	DSM, Netherlands
Vectra	LCP	Hoechst
Vedril	PMMA	Montedison
Vespel	Polyimide	DuPont, USA
Vestamid	Polyamides	Hüls, Germany
Vestan	Saturated polyesters	Bayer, Germany
Vestodur	Saturated polyesters	Hüls, Germany
Vestolen	Apolyethylene	Hüls, Germany
Vestolen P	Polypropylene	Hüls, Germany
Vestolit	Polyvinylchloride	Hüls, Germany
Vestopal	Unsaturated polyesters	Hüls, Germany
Vestoran	SAN	Hüls, Germany
Vestorpen	TPO	Hüls, Germany
Vestyron	PS	Hüls, Germany
Victrex	Polysulfone, Polyethersulfone	ICI, England
Vidar	PVDF	Solvay
Vinidur	Polyvinylchloride	BASF, Germany
Vinnol	Polyvinylchloride	Wacker-Chemie, Germany
Vinoflex	PVC	BASF
Viplast	PVC	Montedison
Viton	Fluor elastomer	DuPont, USA
Vitreosil	Quartz, silica	DuPont, USA
Vitrex	Silicate cement	AtlasMineralProducts, USA
Voltalef	Polytrifluorochloroethylene	Elf Atochem, France
Vulcathene	Polyethylene, low density	—

(Continues)

(Continued)

Trade Name	Chemical Classification	Manufacturer
Vulcoferran	Elastomeric, rubber materials	HarzerAchsenwerke, Germany
Vulkodurit	Elastometric, rubber materials	Keramchemie, Germany
Vulkollan	Polyurethane rubber	Bayer, Germany
Vycor	Quartz/Silica	Corning Glass, USA
Wapex	Epoxy cement	AKZO, Netherlands
Wavistrong	Fibre reinforced plastic piping	FPI, The Netherlands
Welvic	Polyvinylchloride	ICI, England
Xantar	PC	DSM
Xenoy	PC/PBY blend	GEP
Xylon	Polyamides	AKZO, Netherlands
Xyron	Polyphenylene oxide	ASAHI, Japan
Zytel	Polyamides	DuPont, USA

INDEX

B

Printed and bound by CPI Group (UK) Ltd, Croydon, CR0 4YY
03/10/2024
01040434-0007